Fun in Fusion Research

Fun in Fusion Research

John Sheffield

ELSEVIER

AMSTERDAM • BOSTON • HEIDELBERG • LONDON • NEW YORK • OXFORD
PARIS • SAN DIEGO • SAN FRANCISCO • SINGAPORE • SYDNEY • TOKYO

Elsevier
32 Jamestown Road, London NW1 7BY
225 Wyman Street, Waltham, MA 02451, USA

Notices
Knowledge and best practice in this field are constantly changing. As new research and experience broaden our understanding, changes in research methods, professional practices, or medical treatment may become necessary.

Practitioners and researchers must always rely on their own experience and knowledge in evaluating and using any information, methods, compounds, or experiments described herein. In using such information or methods they should be mindful of their own safety and the safety of others, including parties for whom they have a professional responsibility.

To the fullest extent of the law, neither the Publisher nor the authors, contributors, or editors, assume any liability for any injury and/or damage to persons or property as a matter of products liability, negligence or otherwise, or from any use or operation of any methods, products, instructions, or ideas contained in the material herein.

British Library Cataloguing-in-Publication Data
A catalogue record for this book is available from the British Library

Library of Congress Cataloging-in-Publication Data
A catalog record for this book is available from the Library of Congress

ISBN: 978-0-12-407793-5

For information on all Elsevier publications
visit our website at store.elsevier.com

This book has been manufactured using Print On Demand technology. Each copy is produced to order and is limited to black ink. The online version of this book will show color figures where appropriate.

Working together to grow
libraries in developing countries

www.elsevier.com | www.bookaid.org | www.sabre.org

ELSEVIER   BOOK AID International   Sabre Foundation

Transferred to Digital Printing in 2013

Dedication

I dedicate "Fun in Fusion Research"
to my grandson, Gabriel Quinn Sheffield, in the hope that he will
enjoy his career as much as I have mine.

Contents

Acknowledgments

I have written this book to acknowledge the contributions of the occasionally nutty, sometimes egocentric, and truly inventive scientists who have given the world the contraptions, theories, and building blocks that offer hope for the realization of fusion energy. It has been and continues to be a great pleasure to be able to work with so many gifted and entertaining people.

I am deeply indebted to the people who read and gave me sound advice on improving this book: my Atlanta editor, Ann Kempner Fisher; Vicki Kestranek and Carolyn Robbins, the members of the North Point Barnes & Noble and Roswell Library literary critique groups; the Atlanta Writers Club; George Weinstein; Steve Dean; Bertie Robson; Jedwin Smith; and my wife, Dace. I have benefited enormously from what I have learned as a member of the Atlanta Writers Club.

I am also grateful to the people who helped me obtain figures for my memoir: David Campbell, Steven Combs, Don Correll, Martin Cox, Stephen Dean, Eraina Elliott, Tudor Johnston, Martin Laxaback, Dale Meade, Albert Opdenaker, David Rasmussen, Francesco Romanelli, Lynda Seaver, Shihoko Soga, Ronald Stambaugh, and Michael Zarnstorff. I also appreciate the patience and help of Elsevier's Erin Hill-Parks, Lisa Tickner, and Vijayaraj Purushothaman in getting the book improved and published.

Introduction

The fusion of light elements is the source of energy in the sun and other stars. This energy has been released on earth in the laboratory and, dramatically, in the hydrogen bomb.

Why, then, is it the only energy resource that has not yet been deployed for peaceful purposes? The reason is simple: fusion on earth requires a temperature of 100 million °C, which is hotter than the sun's surface.

Fusion researchers have been working at the task of putting this energy to good use for around 70 years. These are exciting times today, but the goal of achieving commercial fusion remains decades away. The creators of the program laid the foundations of our fusion cathedral. The crypt already contains the tombs of many of the major contributors, and many more of us will not see our dream realized.

Today, those who remain are working on the walls and buttresses of the cathedral in preparation for erecting the roof. I've worked in the field for 52 years and counting. So why have we dedicated our lives to fusion energy development, when its realization always seems so far away? Do you have to be crazy? Not really—but it helps.

In an address in the late 1960s, at a conference in Innsbruck, Austria, the eminent Russian fusion scientist Lev Artsimovich offered this parable to explain the mentality of fusion researchers.

Two young men, Sasha and Dima, were wandering around the desert looking for a chance to make a few dinars. Eventually, they came to a lush oasis. They split up to look for opportunities before meeting later in the day to discuss their progress.

"I didn't find much possible work, other than collecting camel droppings for fertilizer," said Dima. "How about you?"

"I found something really interesting," Sasha replied. "I came across a good piece of land that the sheik owns, and it isn't being used."

"So, what did you do?"

"I asked the sheik if he would let us use the land for a year. We could grow crops and make money."

"What did the sheik say?"

"He asked me what I would do for him."

"And . . .?"

"I discovered that the sheik has a donkey that he believes is the smartest animal in the world. He's continually boasting about its prowess to other sheiks."

"What's that got to do with the land?"

"I told him I would teach the donkey to speak French."

"You're crazy." Dima recoiled in horror. "He didn't agree, did he?"

"He agreed," Sasha said, grinning. "But he set a condition. If I don't teach the donkey to speak French within the year, he'll cut my head off."

"You're a dead man, Sasha."

"I don't think so, Dima. You see, there are three reasons I'll keep my head. First, during the year, the sheik may die, and the agreement will be voided; second, the donkey may die; and third, maybe I'll teach the donkey to speak French."

Because this book is designed to show the humorous side of research, I have put the bulk of the discussion of fusion at the end of this book, in Chapter 17, along with a list of acronyms and a glossary of terms. The technical discussions in the body of the book are only designed to introduce the area in which I was working at the time. For a very detailed discussion of fusion energy, I recommend that the reader should look at the 1999 document, prepared for DOE by the Fusion Energy Sciences Advisory Committee, *Opportunities in the Fusion Energy Sciences Program* (http://www.ofes.doe.gov/more_html/FESAC/FES_all.pdf).

My story is not intended to disparage our efforts, nor to give the impression that scientists spend all their time in frivolous activities. On the contrary, most scientists work very long hours, and some occasional light relief makes it easier to bear. I describe the tremendous advances that have been made toward meeting the goals, along with the fun side of research: the wild and wacky people and experiments, the misadventures and wit that I expect could be found in any arena of research and development. Scientific research is not the dry, disciplined arena that many imagine—that's because it involves people, with all of their foibles, fantasies, and frailties. Nevertheless, despite the subject being easy to mock, I remain firmly convinced that fusion energy will become a commercial venture someday.

1 The Fusion Dream

"Endless possibilities... and zero chance of success."
Dr. Walter Marshall, regarding the prospects of fusion energy being realized.

In 1954, the Zero Energy Thermonuclear Assembly (ZETA) came into operation at the United Kingdom Atomic Energy Authority's (UKAEA) laboratory at Harwell, 15 miles south of Oxford. Great excitement followed the announcement in 1958 that it had produced thermonuclear neutrons. The newspapers had a field day with headlines, which the *Daily Mail* won for crassness with "Unto Us a Sun Is Born."

In 1958, I was finishing my bachelor in physics degree in London at Imperial College and took an optional course on fusion energy—the thermonuclear power source of the sun and other stars—and learned about ZETA and other experiments. Fascinated by this exciting area and thrilled to find an alternative option to national service in the military, I joined the thermonuclear division at Harwell. Later tests showed that the ZETA neutrons had *not* been produced by fusion. That's research for you.

At the time I started working in fusion, Dr. Walter Marshall was head of the theory division at the laboratory and subsequently became director. Later, he was knighted and, later still, raised to the peerage as Lord Marshall of Goring. This was a well-deserved award because, as evidenced by the quote at the start of this chapter, he had clearly exhibited the ability to gore.

Although his sarcastic remark was not heartwarming for a young researcher like me, I have retained a fascination with fusion energy and remain convinced that commercial success will be realized in this century.

As I said in the Introduction, scientific research is not the dry, disciplined area that many imagine, which reminds me of something that occurred many years later in Gaithersburg, Maryland, on a dry, chilly day in March. By that time, I was a program leader of fusion work at the Oak Ridge National Laboratory and, along with other program leaders from laboratories and universities, was doing the annual "show and tell" for the Department of Energy (DOE)'s Office of Fusion Energy. A professor from a major university described how an expensive piece of equipment wouldn't fit into an access port when they cooled their experiment down.

"You mean it isn't going to work?" his program manager chided. "You've wasted our money?"

Fun in Fusion Research. DOI: http://dx.doi.org/10.1016/B978-0-12-407793-5.00001-7

The professor reached out toward the projector. "Not a big deal. We can fix it. I'll show you how on the next viewgraph." Suddenly a spark flashed from his finger to the mirror mount. He recoiled and exclaimed, "Does this happen often?"

From the back of the room, a DOE program manager shouted, "Only when you lie."

Big Bang

As I understand it, in the beginning was the Big Bang—the ultimate cosmic sexual experience. Subsequent to the Bang, various forces came into play in our evolving universe—respectively, gravity and the electro weak and strong nuclear forces. Most important, these forces led to fusion, which produced and is still producing the elements today. This fact was first understood when Arthur Eddington proposed that fusion of hydrogen was the source of the sun's energy.

Figure 1.1 illustrates the important realities of fusion material: it requires high temperatures—the surface of the sun is around 5 million °C. It can be turbulent, as seen in the eruptions from the surface. Nevertheless, it can be contained by gravity in the case of the massive sun and by magnetic fields and inertia on earth.

In the fusion process, fundamental particles combine to produce a larger particle (Figure 1.2).

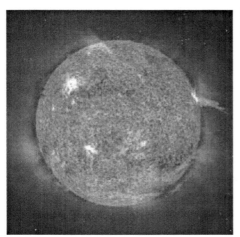

Figure 1.1 Photograph of the sun.
Source: Courtesy of NASA.

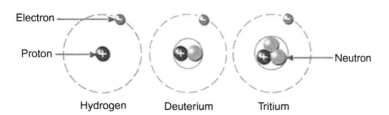

Figure 1.2 The three hydrogen isotopes.
Source: Courtesy of General Atomics.

For example, and of interest on earth, at high temperatures (unfortunately for us 100 million °C or more), the nuclei of heavy hydrogen—deuterium (D) and tritium (T)—combine to make helium and a neutron in D—T fusion. The initial particles weigh more than the products, and the mass difference is released as a kinetic energy of 17.6 million electron volts (MeV). The equivalence of mass and energy was explained by Einstein as energy, $E = mc^2$, where c is the velocity of light and m is the mass (Figure 1.3).

I have always liked the wonderful cartoon (by S. Harris, I believe) that shows Einstein at a blackboard looking at ~~$E = ma^2$~~ ~~$E = mb^2$~~ $E = mc^2$ √.

Interestingly, nearly all of the energy resources available to us on earth come from nuclear fusion in the core of our sun and other stars, sunlight, which gives us plant growth, wind and waves, water evaporation to rain, and then hydroelectric power. Then there's the fission of heavy elements, such as uranium and thorium, which are produced in supernovas, fossil fuels made from elements produced by fusion, and geothermal energy, which arises from radioactive decay in the earth's core.

The evidence for fusion's formidable power has been made clear in hydrogen bomb tests, but it remains the only energy resource available to us that we have not yet exploited for peaceful purposes. While there may be other sources of energy that are realizable, if they exist at all, they would be related to those forces that were important closer to the Big Bang, and at temperatures far higher than 50 billion °C.

It seems improbable that these other sources could be of practical use since even investigating such a region cannot be done efficiently today. Therefore, it's far

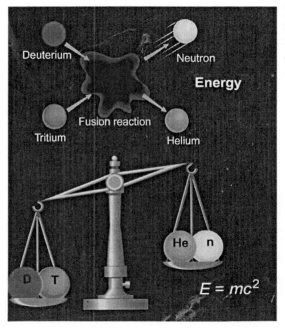

Figure 1.3 Deuterium and tritium nuclei fuse to produce a helium nucleus and a neutron.
Source: Courtesy of Fusion Power Associates.

more sensible to consider only those sources of energy that have been identified as economical or potentially economical, which includes fusion.

This example was given to me by Richard Post of the Lawrence Livermore National Laboratory (LLNL) to illustrate the sheer magnitude of the energy contained in the deuterium, which is one part in 6500 of all the hydrogen on the earth:

> If we were to extract the deuterium from the water that typically flows through an 18-inch water main, and burn it and its reaction products in fusion reactors, it could continuously supply about 2000 billion watts of electricity—the world's total electricity supply in use in the early 2000s.

The heaviest isotope of hydrogen (tritium) is radioactive, and it decays. We have to produce it by bombarding lithium with neutrons, and lithium is not a limitless resource.

Deuterium is, in effect, limitless in the oceans. Therefore, the ideal reaction would be deuterium−deuterium (D−D) fusion, but it requires a temperature of around 400 million °C. Consequently, current research focuses on D−T, despite the additional complication introduced by the requirement for lithium.

There are numerous ways to produce temperatures of 100 million °C or more. For example, passing large currents (millions of amperes) through the fusion fuel, heating it with microwaves, and bombarding it with intense particle or laser beams. Using such techniques, more than 500 million °C has already been achieved in the laboratory.

At temperatures above 10,000 °C, electrons are stripped from atoms and the resulting material has equal numbers of free electrons and positive ions, such as flames, arcs, the sun, neon lights, and most of the visible universe. This state is known as a *plasma*, which is sometimes called the fourth state of matter— solid→liquid→gas→plasma. The changes occur as the temperature is raised (e.g., ice melts to become water) (Figure 1.4).

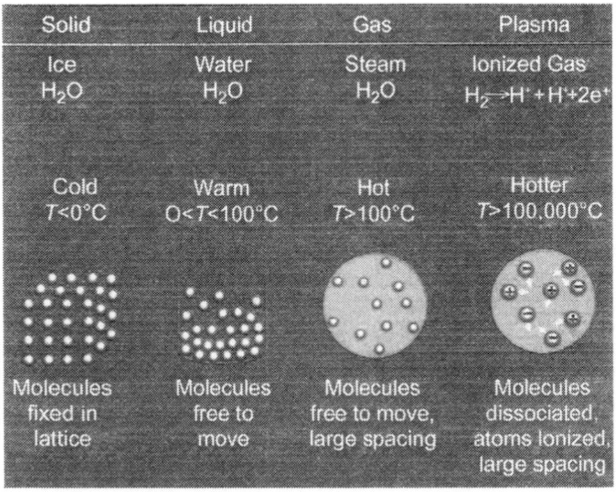

Solid	Liquid	Gas	Plasma
Ice H_2O	Water H_2O	Steam H_2O	Ionized Gas $H_2 \rightarrow H^+ + H^+ + 2e^+$
Cold $T<0°C$	Warm $0<T<100°C$	Hot $T>100°C$	Hotter $T>100,000°C$
Molecules fixed in lattice	Molecules free to move	Molecules free to move, large spacing	Molecules dissociated, atoms ionized, large spacing

Figure 1.4 As the temperature is raised, a solid turns into a liquid, then to a gas, and finally to a plasma. *Source*: Courtesy of General Atomics.

This is the realm of plasma physicists like me. In the fall of 1967, our group hosted the American Physical Society's (APS) annual Plasma Physics and Fusion Research Conference. I chaired one of the more obscure sessions, in which speakers got 12 minutes, including questions. With numerous parallel sessions and people rushing from one session to another to hear topics of interest, keeping on schedule was a challenge. A second challenge was presented by the APS policy that allowed any member to present one of these brief talks. The word *plasma* has more than one meaning, and the audience in my session listened in bewilderment as an earnest young man talked about fibrillation in the hearts of horses—blood plasmas.

A Cutting Review

Temperatures such as 100 million °C are hard for most people to comprehend. An obvious question is: How can you measure them? One way is to shine a laser beam into the plasma. Some of the laser light is scattered by the hot electrons and shifted in wavelength because of the electrons' motion—a Doppler effect.

For those who remember the time when people were allowed to smoke in movie theaters think about the way in which the projected light could be seen scattering off the smoke particles above your heads. The more smoke, the brighter the light appeared. If it had been possible to actually analyze that light, you would have seen a subtle change in wavelength caused by the motion of the smoke particles through the projected light. In the case of plasma experiments, the scattered light contains information about the electron density and velocity distribution (i.e., temperature). A bigger spread in wavelength results from a higher temperature. At 100 million °C, red light will be shifted down to the blue part of the spectrum. The advent of powerful, single-wavelength lasers brought the scattered signal into the range where the spectrum of scattered light could be measured.

In 1975, I published a book on this subject. I remain proud of this accomplishment, but I found out that not everybody shared my enthusiasm for the book when, in the early 2000s, I attended an enjoyable meeting in Oxford. At the reception, Paddy Barr told me the following story about my book. Years earlier, he had worked in a university in North Wales, where he was engaged with his group in the scattering of radiation from plasmas. My book on scattering, which was in their library, was used as a key reference.

Among the students was a man, an ardent Welsh nationalist, with a *different* approach to life. An example of his awkward behavior was his insistence that his examinations be given in Welsh—which was his right, nevertheless. That meant that Paddy had to send the exam to South Wales to get it translated. The student would then answer the exam questions in Welsh, and the answers were sent back to South Wales for translation into English so that examiners could grade him; a pain in the butt for all involved.

One day, Paddy went into the library to check on something in my book. To his horror, he found the book in a sorry state. It appeared that someone had stabbed it repeatedly with a large kitchen knife! After checking around, Paddy and his colleagues concluded that the student had committed the act. In fact, the man readily admitted that he had, but he never explained why.

It was then that Paddy insisted that the student put transparent sticky tape across the cuts on each page. This created a monstrously thick version that fanned out from the binding. When the group ceased to exist, Paddy moved to another job and took the book with him. Decades later, it is still that weird shape and all of the tape has turned yellow.

I was disappointed to hear that my book received such a cutting review, but better my book than me.

Genesis of Fusion Research

As previously mentioned, fusion remains the only energy resource that we have not yet used for peaceful purposes. Research on fusion energy started during the 1930s and 1940s in Europe, the then-Soviet Union, and the United States. The challenge was obvious—*how to produce hot fusion fuel at 100 million °C and then contain it long enough for it to fuse and generate more energy than the amount used to create it.*

Three ways in which a plasma may be contained are using gravity, as in the sun, using magnetic fields, and using inertia, as is done in the hydrogen bomb and inertial fusion (Figure 1.5).

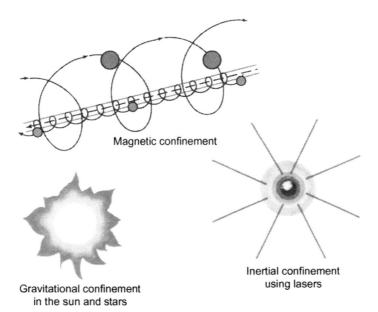

Figure 1.5 Three ways to contain high-temperature plasmas.
Source: Courtesy of General Atomics.

Magnetic Fusion

The early efforts to produce fusion plasmas used cylindrical systems. The first such approach was tried in the 1930s by Arthur Kantrowitz in a laboratory at the National Advisory Committee for Aeronautics (NACA)—the forerunner of NASA, the National Aeronautics and Space Administration. A patent application was submitted in 1941.

It was realized that such systems suffered badly from end losses, even when very high magnetic fields at the ends constricted the loss region, called a *magnetic mirror*. Nevertheless, such systems led to the development of fascinating and amusing science.

Around 1960, Bertie Robson, later a colleague of mine at the University of Texas, had a mirror experiment housed behind a simple fence in a hangar at the Harwell laboratory in rural Berkshire in the United Kingdom, a county in which Morris dancers in medieval costumes still celebrate May Day. The experiment used a series of copper coils to produce a magnetic field.

One day, a cleaning lady was mopping the concrete floor by the fence. The moment she picked up her bucket of water, the operator switched on the magnetic field. The metal bucket, with the cleaning lady holding on, was dragged up toward the magnet. She dropped the bucket and fled, screaming "Witchcraft!"

No one was able to persuade her to return to the "evil building."

In the 1960s, a combination of a magnetic mirror with energetic electron beams was constructed at LLNL by Nicholas Christofilos. The goal was for the electron beams to reverse the field and create a closed field-line system, thereby eliminating the end losses. After many years of continual additions and adjustments by Christofilos to fix a series of problems, two scientists were discussing the strange evolution of the device.

"Nick's a brilliant man," said one scientist. "What's the problem?"

"Oh, it's simple," the second scientist replied. "On Nick's planet, all the electricity is produced by Astrons. Unfortunately, Nick can't quite remember how they work." In the early 1970s, this system was abandoned.

The main approach used to eliminate the problem with the ends of linear systems is to bend the configuration into a circle inside a toroidal vacuum vessel. Examples of a torus are a tire and a doughnut (Figure 1.6). I apologize for introducing some technical jargon, but it clarifies later discussions. *Toroidal* refers to the long way around the torus, and *poloidal* to the short way around, as illustrated in the figure 17.3. A glossary of terms is at the end of the book.

In the earliest toroidal experiments, the magnetic field was provided by driving a toroidal current that also produced and heated the plasma—the *diffuse pinch*. In the late 1940s, Sir George Thomson patented a diffuse pinch fusion reactor in Britain, using magnetic fields to contain the hot fuel. His patent contained many of the features found in today's design studies of fusion power plants. Sir George was the son of Sir J.J. Thomson, discoverer of the electron, who was known as "father of the electron." The epithet makes you wonder what Sir George looked like.

Magnetic confinement

Fast-moving particles in a simple container would quickly strike the walls, giving up their energy before fusing

Magnetic fields exert forces that can inhibit and direct the motion of the particles

Magnetic fields can be fashioned into complex configurations sometimes called magnetic bottles

Figure 1.6 Magnetic fusion energy (MFE) uses magnetic fields, which have the property that the electrons and ions in the plasma spiral around the field lines and, in principle, may be isolated from material walls.
Source: Courtesy of Fusion Power Associates.

Unfortunately, such systems, which only have the field owing to the current (see Figure 7.10), are unstable and, in the 1950s, the following occurred.

In Russia, Igor Tamm and Andrei Sakharov added a large toroidal field—the *tokamak*. The combination of the field from the plasma current and the field from the toroidal coils leads to a total field that spirals around the torus and is very effective at containing the plasma.

In Harwell, England, Roy Bickerton suggested adding a weaker toroidal field, leading to the *reversed field pinch*, so called because the magnetic field spontaneously reverses, leading to a system with improved confinement, as discovered at ZETA. This approach was also pioneered at the Los Alamos National Laboratory.

The third proposal came from Lyman Spitzer at Princeton, who proposed using twisted external coils to produce such a twisted field—the *stellarator*.

Success in the world's tokamak program led to the start of construction of the International Thermonuclear Experimental Reactor (ITER) in the south of France. This project is a collaborative effort of China, Europe, India, Japan, South Korea, Russia, and the United States. ITER is designed to produce 400 MW of fusion power—10 times the power used to heat the plasma.

Inertial Fusion

The dramatic explosions of hydrogen bombs is a proof that it is possible to release fusion energy using inertia to hold a compressed and heated capsule of fusion fuel together long enough for substantial fusion to occur (Figure 1.7).

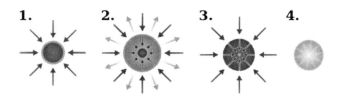

Figure 1.7 Schematic of the stages of inertial confinement fusion using lasers:
1. Laser beams or laser-produced X-rays rapidly heat the surface of the fusion target, forming a surrounding plasma envelope.
2. Fuel is compressed by the rocket-like blow-off of the hot surface material.
3. During the final part of the capsule implosion, the fuel core reaches 20 times the density of lead and ignites at 100,000,000°C.
4. Thermonuclear burn spreads rapidly through the compressed fuel, yielding many times the input energy. The blue arrows represent radiation, orange is blow-off, and purple is inwardly transported thermal energy.
Source: Courtesy of LLNL. (For interpretation of the references to color in this figure legend, the reader is referred to the web version of this book.)

The attempts to make inertial fusion energy (IFE) systems for peaceful uses started with the development of powerful lasers in the 1960s, an effort led by Nikolay Basov in the Soviet Union and John Nuckolls at LLNL. The first experiment at LLNL consisted of 12 beams of a ruby laser called 4-pi, aimed at the center of a spherical vacuum vessel. The larger Shiva laser with 10 kJ of laser light succeeded 4-pi, and in turn, that was replaced by the Nova laser, which delivered 40 kJ.

Since that time, dramatic progress has been made, and the National Ignition Facility (NIF), the newest laser at LLNL, has 192 beams and can deliver up to 2 MJ (in energy, equal to what is produced by about a pound of high explosives) of ultraviolet light energy at 3500 Å. It is designed to produce about 20 MJ of fusion energy from a single capsule. NIF is now operating and producing very encouraging results.

The production of tiny spherical capsules (typically, 1/16th to 1/8th of an inch in diameter) is a remarkable feat. The requirement is for a perfectly spherical shell of solid deuterium and tritium to have a surface roughness not to exceed something like one part in a million. The reason for this precision is that any imperfection can get amplified by instabilities during the compression, leading to mixing of the shell and fusion material and, when bad enough, very little fusion. The problem of instabilities in all fusion approaches is discussed in Chapter 17.

Now hydrogen isotopes don't become liquid until the temperature drops close to absolute zero (−276°C). One of the approaches to achieve this goal is to use a thin spherical shell of porous plastic. This shell is immersed in liquid deuterium and hydrogen and wicks it up into the pores. X-ray images of the shell generally show that the thickness of hydrogen is not uniform, which is unacceptable for subsequent good compression and heating. However, tritium is radioactive and decays, providing a small amount of heat to the shell. After an hour or so, the tritium and deuterium on the inside of the shell have sublimed, evening out the thickness.

Today, shells of various kinds are produced by General Atomics in San Diego, but when the research started, the challenge was to find a producer of perfect shells. Some bright scientist realized that a Swedish company was making enormous numbers of tiny spherical polystyrene shells for use as insulation. He discovered that by careful selection from millions of these shells, one could obtain a few that met the requirements of inertial fusion.

In 1956, Sebastian (Bas) Pease, later director of the UKAEA's Culham Laboratory, quipped, "Our vision is of a power station, sited perhaps on the coast, with a pipe bringing water from the sea, helium leaving the chimney, and electrical power flowing into the grid. We do not know what to put inside the power station."

In fact, it would not be until the end of the 1960s that fusion would show real promise. That was when the Russian tokamak, T-3, achieved an electron temperature of 10 million °C (1 keV). Since then, the temperature has been raised to more than 650 million °C.

What is the reality of fusion energy? In 1998, I organized a workshop for the DOE's Office of Fusion Energy Sciences (DOE-OFES), in which 11 fusion experts met with 6 nonfusion representatives, all knowledgeable about environmental and general energy issues. They were from the U.S. Public Interest Research Group, Union of Concerned Scientists, Public Citizen, Natural Resources Defense Council, the University of Tennessee's Joint Institute for Energy, and the JFK School of Government at Harvard University.

The purpose of the meeting was to discuss their views of fusion as an energy source. At the beginning of the meeting, each attendee was handed a list of questions to answer anonymously, except there was a color code to distinguish between fusion and nonfusion representatives. Interestingly, while some of the environmentalists were totally opposed to anything nuclear, there was a little difference in the two groups' understanding of the strengths and weaknesses of fusion energy.

Other than the smart remarks—"great dinner table discussion (topic)" and "like a religion"—the positives included *Abundant and clean energy source—no CO_2 and could be better than fission; and interesting science and spin-offs.* The negatives included *Cost and timescale, complexity; and generates radioactivity.* And there was one unresolved disagreement about the value of IFE research for peaceful purposes being of value to people who wanted to build a hydrogen bomb.

The next decades will take us from the results of NIF and ITER even closer to our ultimate goal of economic energy production. However, despite the remarkable progress illustrated in Figure 1.8, I agree with the views expressed above for, in 2013, fusion energy, despite substantial progress, is still not ready to make economical electricity. As I mentioned in the Introduction, the timescale for development, approaching 100 years, is similar to that required to complete the great cathedrals of medieval Europe. I presume that faith, and the fact that such edifices grew from year to year, kept those early builders going. A steady stream of successes and the dream that they can harness this energy source for peaceful uses continue to drive fusion researchers in the second half of their 100-year effort.

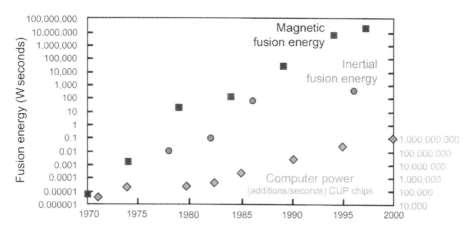

Figure 1.8 The rate of progress in fusion energy production exceeds that in computer power.

Source: Courtesy of Dale Meade, PPPL.

I occasionally hear a pundit state that either magnetic or inertial fusion is the way to go, and the other approach is nuts. I don't doubt that net energy will be produced by both approaches. However, both are incredibly complex, and the main challenge, in my view, is to come up with systems that have the reliability, maintainability, and availability to be a commercial power plant. With regard to choosing between them, today, I am reminded of Woody Allen's comment, "More than any time in history, mankind faces a crossroads. One path leads to despair and utter hopelessness, the other to total extinction. Let us pray that we have the wisdom to choose correctly."

2 Harwell Heydays

Arcs and Sparks

A major feature of fusion research is the use of large amounts of electricity—very high currents and voltages. Think lightning. This was the area in which I started my career.

When I joined the Atomic Energy Research Establishment at Harwell in 1958, there were three groups in the thermonuclear division, led by Roy Bickerton in Building 488, Sebastian "Bas" Pease in Hangar 7, and Peter Thonemann in Building 150. In addition, research was undertaken at the weapons research center at Aldermaston and at Associated Electrical Industries (AEI)-Aldermaston.

At Harwell, I was an assistant experimental officer—think glorified technician—assigned to work for Jim Paul on the Spider shock wave experiment in Building 488. The building consisted of two high-bay experimental areas with overhead cranes. Offices and small labs were spread around three sides. Each bay held a number of experiments.

Our experiment's name was a recognition of the spiderlike appearance of all the high-voltage (HV) cables that connected the capacitors and their switches to the metal shell segments that encased its quartz torus (Figure 2.1).

My first job was working on its 20,000-V switching systems. The switches, which could close in a small fraction of a microsecond (i.e., a millionth of a second), were used to discharge the energy of capacitor banks to drive large currents in a plasma. In turn, these currents heated the plasmas. The main heating approach in Magnetic Fusion Energy (MFE) remained driving large currents in the plasmas. This was a cheap and effective way to make plasmas, but one that suffered from the problem that as plasma became hotter, its resistance dropped, and the heating became successively less effective.

In shock wave experiments, the current was raised very rapidly, i.e., in less than a microsecond, causing the plasma to pinch inward and creating shock waves that gave additional heating. The purposes of Spider were twofold: Undertake fundamental plasma physics studies, comparing theory and experiment, and find out if shock waves would allow plasmas to attain high temperatures.

The energy in our capacitor banks ranged up to 1 MJ, equivalent to about half a pound of high explosives—*hand grenade-like*! Therefore, the UKAEA required that we employ stringent safety precautions to protect ourselves from HV and explosive hazards. Each experiment was housed inside a 10-ft-tall fence—solid metal for 3 or 4 ft, and metal-framed wire mesh above it. In some cases, sheets of

Fun in Fusion Research. DOI: http://dx.doi.org/10.1016/B978-0-12-407793-5.00002-9

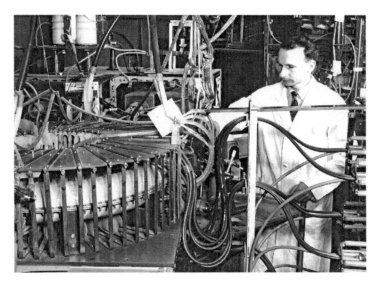

Figure 2.1 Jim Paul stands by Spider, circa 1959. A section of the toroidal copper shell shows through a set of copper bars that produces an axial magnetic field. HV cables provided current from capacitor banks to the four segments of the shell.
Source: Photograph reproduced courtesy of UK Atomic Energy Authority.

clear Lucite were mounted inside the mesh to give additional protection against flying objects.

Nevertheless, not everyone was as careful as he should have been. One notable experimental officer was a small man who was by turns, cocky or surly and generally lazy. We called him "Thundering Sam" for the delight he took in blowing things up. On one occasion, a colleague, Roger White, and I were walking past an HV cage. The door was open, and we saw Sam sitting on a chair cutting through a 1 in diameter, HV cable, which was connected to the switch he was testing. Sam was giggling. Like all HV equipment, Sam's 40 kV (half a megajoule) capacitor bank was protected by automatic mechanical and electrical systems to discharge the HV if the cage door were opened, i.e., connect the HV components to the Earth. In addition, it was the rule to attach a metal grounding stick on a long insulated handle to the exposed conductors to confirm that the safety systems had worked.

"What's funny, Sam?" Roger asked.

"Look at the sparks coming from the saw blade," Sam replied.

"You silly bugger, the voltage is still on," we screamed, realizing that Sam had cut through the outer plastic cover of the cable and into the copper braid that formed the return path for the current. The saw blade was only eighths of an inch away from the central HV conductor. If Sam cut through a fraction more of the polyethylene insulator, the resulting arc would have blown up the saw blade in his face.

We rushed into the room and pulled Sam through the door.

"Did you hand ground it?" asked Roger.

Sam, looking a bit shaken, said, "No need, the systems are automatic."

I took a grounding stick and cautiously shoved the tip toward the nearest conductor. The noise of the ensuing arc echoed around the high-bay area.

Roger, who'd been studying the mechanical grounding system, pointed at a link hanging down. "System's broken, Sam."

"But the electrical one should have worked," Sam replied defensively.

"Connected to the other side of the link," I said. "Why were you cutting the cable?"

"Otherwise I'd have had to undo all those bolts," said Sam, indicating the array holding the switch together.

"Better do it," Roger said.

Sam was still muttering about the work he now had to do when Roger and I left.

During this period, another colleague, Syd Hamburger, took a sabbatical at the University of Oklahoma, where the safety precautions were less stringent. He stopped his experiment to work on a capacitor bank, not realizing that there was no automatic grounding system. The good news is that when he touched a live conductor, he was thrown violently away. (At lower voltages, one can become stuck to a live conductor. We were trained to use an insulator—a towel, or a piece of plastic—to pull someone off if this occurred.) The bad news is that Syd fell back onto a second live conductor that discharged a capacitor bank between the nape of his neck and his butt. Luckily, he only lost a little hair, but was otherwise okay.

While I was engaged in maintaining the HV switches on the Spider experiment in Building 488, John Martin worked on another facility in our high-bay area. The two experiments were separated only by one of the corridors that gave access between the metal enclosures. It had taken John a whole week to painstakingly wind layers of 8 in diameter, insulated copper wire onto a toroidal vacuum vessel. This solenoid would produce a toroidal magnetic field when a large current was passed through it. When the day came to fire up this device, my colleagues and I were in the metal screened control room for Spider. We were fortunate to have this protection, for suddenly a loud bang was followed by a sustained tinkling sound. John's copper coil had exploded. It broke on the inner median plane where the magnetic forces were highest, creating hundreds of 1-ft-long, thin copper arrows that rained all over the bay. At that point, I knew how the English must have felt at the Battle of Hastings when the Norman archers started firing away. I guarantee that if Building 488 still stands, there are a few copper missiles still stuck in the ceiling.

Located next to the domain of Thundering Sam was an experiment called the Un-pinch, another approach to containing hot plasma. It consisted of a 15-ft-long Pyrex cylinder with an insulated, 2-in-diameter brass rod down the center. One end of the rod was connected to a thick, circular brass plate. At the other end, the insulated rod passed through a hole in a second brass plate. The plates closed off the vacuum system and acted as electrodes. This whole contraption was mounted on the floor above a large capacitor bank. Across the corridor from the Un-pinch was a delicate experiment in which Ian Spalding was measuring instabilities when current was driven in a column of mercury. I shared the lab/office under Ian's experiment with my colleagues Mike Kasha and Tony Malein. One day, a huge bang and clattering noises from the direction of the Un-pinch drew us into the corridor. We saw that the protective fence above us at the end of the Un-pinch and the fence "protecting" Ian's experiment had been demolished by the monstrous brass discus that had been hurled off the Un-pinch due to inadequate constraints to handle the magnetic forces when an arc occurred.

What I find astonishing is that, in all these explosive events that I saw or heard about, no one was, in fact, killed.

While I enjoyed my job and was learning a lot about real experimental work, I realized that I wanted to be more than a technician. Unfortunately, my degree grade was not sufficient to become a scientific officer, the parallel and superior path for scientists. Fortunately, Jim Paul and Roy Bickerton were incredibly kind and decent people. When Mike Kasha found out that it was possible to do a master's degree part-time at Northern Polytechnic in London, they agreed to fund Mike and me and to let us have half a day a week free to do the coursework. The courses were on electrical discharge physics—arcs, neon lights, and flames—and molecular spectroscopy.

One of Mike's flatmates (I'll call him Rupert) *worked* in another division. He had very little to do and was bored. To while away the time, Rupert played practical jokes. A favorite source of material for his games was the weekly Harwell newsletter, which had a "buy and sell" section. On one occasion, he found an advertisement offering an old bull-nosed Morris sedan. He called up and asked if the vehicle was the rare, amphibious version developed for D-Day, and he persuaded the owner to go out into the parking lot and check if there were little fins on the brake drums.

Harwell was built in the early 1940s on an airfield. By the time I arrived, the main runway still served as a major access road outside the security fence on the south side of the site. Tony Malein, who had joined Harwell some years earlier, told me this story about a US airman flying overhead in a Sabre jet with intermittent engine trouble. To him, the Harwell access road must have looked God-given. Miraculously, he succeeded in landing as his engine gave out, the plane skidding into a wet, grassy area at the west end of the runway, and stopping before he hit a building. Tony was among the crowd gathered by the fence to see the jet, nose down in the mud and surrounded by lab security. The pilot, nonchalantly leaning against the fuselage, was gesticulating wildly.

The lab director called the pilot's base, and a team came out to decide how to remove the Sabre. Apparently, there was a hell of an argument between those who wanted to dismantle the plane and ship the pieces out by truck, and those who wanted to fly it out. In fact, the plane was muddy but hardly damaged, and the problem with the engine was easily fixable. The latter camp won, but there was a problem—the runway was too short for the plane to takeoff under its own power.

The paved length of the runway was about 800 yards. A high, chain-link fence, mounted on concrete posts, extended alongside it. Numerous buildings bordered the north side of the fence, including a small research nuclear reactor, and also a large hole that had been dug for the foundations of an accelerator. A second set of arguments took place between the lab and the US Air Force about the wisdom of trying to fly the plane so close to hazardous areas. The Air Force general in charge of the operation pointed out that the reactor was near the beginning of the takeoff. The plane, traveling from the west, would not be moving fast or be airborne until much nearer the east end of the runway, away from the hazards.

When the big day came, hundreds of employees, including Tony, lined the fence to watch the spectacle. The Air Force had overcome the problem of the short runway by attaching two jet-assisted takeoff (JATO) rockets under the wings to give the needed extra thrust. The pilot arrived in a Jeep, dismounted, pushed his cap back, and inspected his plane, playing up to the crowd.

He climbed into the cockpit and closed the canopy ... silence ... until, with a roar, the engine started. Then, with the engine running flat out, he released the brakes and fired the rockets. The onlookers covered their ears to drown out the deafening noise. Dust and grass flew into the air as the Sabre jet shot forward noisily, belching smoke from under its belly. As the nose wheel rose, the JATO engine on the near side fell off, hit the runway, shot 1000 ft into the air, and performed somersaults. The huge crowd backed away rapidly from the fence.

The pilot was unable to control the unbalanced thrust, and the plane, still with its main wheels on the ground, took a violent left turn, hit the fence at an angle, and headed straight for the accelerator hole. One by one, the fenceposts popped out of the ground, and the plane, enmeshed in chain link, pulled fence and posts toward the hole. By now, the other JATO rocket was exhausted and the pilot had managed to cut off the jet engine. The monstrous version of the arrestor system on an aircraft carrier brought the plane to its knees only feet from the chasm.

A long stunned silence was interrupted by the sound of emergency vehicles rushing to the scene. The canopy lifted, and the pilot flopped out of the cockpit, leaning against the fuselage for support. He pushed back his cap, pulled a bent cigar from an upper pocket, and put it unlit in his mouth. Tony said that from his stance, he reckoned the pilot was thinking, *Was that fun or what?*

3 Stick with Spiders—Tarantula

During my first 2 years at Harwell, I also worked on making diagnostics for the Spider plasmas. Here, my shortsightedness was an advantage, for I could easily wind the 1-mm-diameter, 10-turn coils that we used to measure the rapidly changing magnetic fields in the shock waves. My skill at making these probes compensated (I hoped) for my clumsiness, when I succeeded in breaking one of Spider's delicate and expensive quartz toruses.

At that time, diagnostics for high-temperature plasmas were very limited in scope. We were able to put the material probes into Spider because its plasmas existed for only millionths of a second—too short a time to burn up the probe and for ablated probe material to contaminate the plasma. In contrast, the large toroidal experiment (ZETA), up the road in Hangar 7, had plasma currents of hundreds of kiloamperes for thousandths of a second. Probing the depths of its 1-m-diameter plasma required noninvasive diagnostic techniques, which was the reason that it had been difficult to identify the source of the so-called thermonuclear neutrons, as mentioned in Chapter 1. Spectroscopy—analyzing the light emitted by the plasmas—was a major diagnostic technique, but, like measuring the neutrons, it was sometimes difficult to interpret. During the 1960s, the advent of lasers and the use of scattering of light dramatically improved our capacity to make measurements.

With the exception of ZETA, the experiments on fusion and plasmas at Harwell were modest in scale. The Mark-4 torus, a forerunner to ZETA, occupied an area on the other side of Spider from John Martin's experiment. Sydney Hamburger—hair grown back on the nape of his neck following his accident in Oklahoma—operated this experiment. As he came in one morning, he was greeted with the words, "Sorry about the fire, Syd."

Syd replied, "Ha, ha. Very funny."

Then, as he proceeded to his office, another colleague said, "Hope the flood didn't damage anything." Syd shook his head, wondering why people had chosen that morning to pull his leg, entered his office, and started working on a paper.

After a while of it nagging at him, Syd simply had to go and look. To his horror, he found his experiment surrounded by a large puddle of water. Following a short circuit overnight, the plastic-covered copper wires supplying the coils had caught fire and burned up the whole facility. The aluminum torus looked like a Salvador Dali watch, and only the charred, fire-proofed, wooden support structure remained standing. The automatic sprinkler system had saved the rest of the building.

Fun in Fusion Research. DOI: http://dx.doi.org/10.1016/B978-0-12-407793-5.00003-0

Despite having a reputation for bad luck in his laboratories—on one occasion, a heavy safety sign fell and just missed his head—Syd was a generous man, a pleasure to work with, and he attracted many capable coworkers.

* * *

In the late 1950s, the UKAEA decided to build a new site combining all of its fusion groups from Harwell and Aldermaston. The site chosen was an abandoned naval air station at Culham, 10 miles northeast of Harwell. (I never found out why the Navy would construct a base in a place as far from the sea as it is possible to get in the British Isles.)

Culham was designed to house the follow-up experiment to ZETA—the Intermediate Current Stability Experiment (ICSE)—in a high-bay building (D-1) sited on a 1-acre area of land. A central building containing offices and small labs ran south from D-1 for a quarter of a mile. Other high-bay areas, rooms for power supplies, offices, and labs were attached to each side of the central corridor.

ICSE was designed with a ceramic torus, some 6 m in major diameter and 1 m in minor diameter. It was to be powered by 10 MJ of 100 kV capacitors, driving megaamperes of current in its plasma. Thundering Sam was one of a number of people developing the switches for this huge capacitor bank.

The theory of fusion plasmas was by then quite well developed, except that resistive effects were not well accounted for. ICSE was based on an ideal theory of how a conducting fluid (read this as "plasma" in this case) should behave in a magnetic field—magnetohydrodynamics (MHD). When the head of the UKAEA, Sir William Penny, heard that the basis for the belief that the experiment would reach high temperatures was a theory that did not include the resistance in the plasma, he wisely canceled the project. Later, Paul Rebut, a brilliant young French engineer/scientist, gave the first version of resistive MHD, but his 1963 talk at a conference in Innsbruck, Austria, was not given much attention. A seminal paper written a year later by Harold Furth, John Killeen, and Marshall Rosenbluth is often cited as a vital source on this subject. I first saw the ebullient Marshall Rosenbluth when he gave a talk at Harwell, amusing his audience by defining "to be a theoretical physicist" as follows:

I am a genius.
You are obscure.
He is a plagiarist.

Tarantula

Tarantula was our group's proposal for a follow-up to Spider. Envisaged as a toroidal shock-wave experiment with a 1 MJ, 100 kV capacitor bank, it was also viewed as too ambitious. Dave Ashby, a peer of Jim Paul's in our group, suggested making a 1-m-long cylindrical plasma using a tenth of the original megajoule bank. I was assigned to work with the engineers on the 100 kV switching system. I later found out that it was Dave who had owned the bull-nosed Morris with the imaginary fins on the brake drums.

One day, I asked Jim Paul what "Tarantula" stood for. His answer was that it was merely another spider. I remember suggesting that it might be an acronym for *T*oroidal *A*ssembly *R*unning *A*t *N*ormous *T*emperatures *U*nder *L*ousy *A*dministration. Jim shook his head. But the Tarantula did look like a giant spider, with its 40, 6-ft-long, thick black cables radiating from the cylindrical body of the shock tube—each cable connected to a 100 kV switch that was plugged into a 6-ft-long cylindrical capacitor. The 40 black trigger cables dangling from the switches definitely enhanced this image.

In the winter of 1960, Mike Kasha and I worked part of the time at Culham. We were assigned to help assemble the Tarantula in Building D-1, which, at the time, had a roof but no walls. That winter was bitterly cold, with icy winds and occasional snow blowing across the open airfield. Our kindly chief engineer, Bert Bishop, brought in grain dryers—electrical heating elements with a large fan—to keep us warm.

Some comic relief was provided by a visit of senior Soviet scientists to Harwell, who gave a series of talks. The first speaker was an eminent Russian, Academician Leontovich. We had all struggled through his seminal textbook on plasma physics. Despite his ill-fitting clothes and scruffy shoes, he had enormous presence.

The academician lectured in Russian, had hard-to-see slides covered in equations, and used an interpreter. His talk was difficult to follow, but, for me, a young scientist, it was exciting to be in the presence of such a great man.

At the end of his talk, the interpreter indicated that he would accept questions.

"I have one," said an American colleague, who then proceeded to launch into a multipart, rambling set of comments and questions that left even the English-speaking audience confused. The interpreter struggled to translate. Leontovich listened with a stony face. At the end of the interpretation, he shook his head and said something to the interpreter.

The American made the mistake of asking, "Did you understand my questions?"

The interpreter started to translate. The academician waved at him irritably and surprised us by answering in perfect English. "I understood the meaning of your words but not the sense."

The interpreter, flying on autopilot, translated this into Russian.

Static Electricity

The Spider and Tarantula experiments allowed us to study shock waves propagating in plasmas. Entertaining effects, including shocks of a different kind, may also be obtained from static electricity. In this regard, small Van de Graaf electrostatic generators are great fun. I expect most of us have seen pictures of a woman with long hair sitting on a wooden stool and touching the top aluminum sphere when the machine is cranked up. As the electrostatic charge builds up on her body, her hair separates and rises. I have often used the generator as a prop in talks on fusion research, to illustrate electrostatic charge repulsion.

Later, when I taught in the physics department at the University of Texas, I would ask for two student volunteers without telling them what they were volunteering for. I would place a metal tray on the sphere and fill it with Rice Krispies. The cereal would shake for a brief period before exploding out 10 ft or so in all directions. I would then produce the dustpan and brush that I had hidden in the lectern and hand them to the unsuspecting janitors to clean up the mess.

Dress Code

In the 1950s and 1960s, it was quite commonplace for scientists to wear a suit to work, or at least a sports jacket and tie. We experimentalists wore white lab coats, issued by the UKAEA, a necessary protection for our own clothing. The UKAEA also provided a variety of protective shoes and boots with a steel toe-cap under the leather. When Mike Kasha and I, poor assistant experimental officers, realized that one of the brown boots was quite a fashionable design, we rapidly obtained some. I wished that I had done it earlier. It would have saved me the agony I felt when I dropped a 200 ft^3 gas cylinder on my right foot. The site nurse took a tiny drill and made a hole through my big toenail to relieve the pressure of blood buildup. Amazingly, this doesn't hurt ... unless, of course, the drill goes too far.

The steel-toed boots were a little heavy, but we soon forgot we were wearing them. One day, Mike, frustrated when an automatic ticket machine in the London Underground took his money but failed to deliver a ticket, gave what was supposed to be a gentle kick to the side of the machine and buckled the metal panel. Turning, he saw the alarmed faces of the people in the queue behind him as they backed away.

On the subject of attire, I am reminded of the time that Emlyn, a draftsman who worked with our engineering group (and a walking example of the word *timid*), made an announcement during our coffee break.

"I'm going to learn to drive," he said. "You all drive. It isn't that difficult, is it?"

"Piece of cake," replied our chief engineer, Keith Plummer. "You can ride a bike, can't you?"

"I guess," said Emlyn, looking uncertain. "Tell you what," Keith said. "I'm going into town at lunch. You can come with me, and I'll show you how easy it is."

Emlyn eyed him nervously. Keith had a sporty, stick-shift sedan and a reputation for driving fast. "I suppose I might."

"Good, that's settled. See you in the parking lot at twelve."

At one-thirty, they returned. Keith's right trouser leg was torn, and Emlyn was even more pasty-faced than usual. I waited until Keith had gone before collaring Emlyn. "How did it go?"

Emlyn shook his head. "I'm not sure about this driving business."

"What do you mean?"

"Keith showed me how to change gears, signal, turn, and so on. Everything was fine on the way out. But on the way back." Emlyn stopped and took a deep breath.

"You know that bend about a mile out of town, the narrow one with a stone wall on one side?"

"Sure."

"Keith drives awful fast. I was keeping an eye on the speed. We were coming into the bend at sixty. 'I'll show you how easy it is,' he said."

"What did Keith do?"

"He put his feet up on the dash. 'Ta da. See, driving's easy. Now all I have to do is drop my feet and change to a lower gear. We'll slow down.'" Emlyn shook his head again. "After Keith put his feet down, he went to change gears. He couldn't."

"Why not?"

"Thank God there weren't any other cars. We went around the corner on two wheels. The gear stick had gone up his trouser leg."

As far as I can remember, Emlyn was still not driving by the time I left Culham.

Ray Hardcastle was one of many effective senior experimentalists. I admired his ability to avoid getting his clothes messed up while working ... unlike me. We reckoned he could change the oil in his car wearing a dinner jacket and go straight to a reception with no problem. Imagine my surprise when, as I strolled past his lab on the way to my neutral beam test stand, I found him sawing away on a piece of wood. I did a double take and backtracked to the door.

"Sawing, Raymond?"

He did not look up or say anything. But, as I continued down the corridor, I heard his words wafting after me.

"And Christ too was a carpenter."

4 Ph.D. Experiment and Security

My Ph.D. Experiment and Other Plasmas

Progress in fusion energy development was slow during the early 1960s, in the sense that increases in the plasma performance, commonly characterized by the fusion product *density* × *temperature* × *confinement time*, were not noteworthy (see page 133 confining the plasma energy in Chapter 17). The main progress was in diagnostic development and in fundamental plasma physics connecting theory to experiment.

In this regard, our Tarantula experiment operated well and produced well-defined shock waves in which the temperature jumped from 1 eV (10,000°C) to 200 eV (2 million °C) in a distance of as little as 2 mm.

After Mike and I had completed our master's degrees at Northern Polytechnic in 1962, Roy Bickerton and Jim Paul supported my proposal to get a Ph.D. degree part-time. Roy's Ph.D. topic in 1954 had been the effect of a longitudinal magnetic field on the central part (positive column) of a glow discharge—commonly referred to as a *neon light*. He showed that, in agreement with theory, the voltage required to drive the current decreased as the magnetic field increased because the field reduced the loss of electrons to the wall. Roy just missed discovering the fascinating result observed by Bo Lehnert in 1955 in Sweden that, at a higher critical magnetic field, the driving voltage increased again. This phenomenon was explained theoretically in 1960 by Boris Kadomstev and Arthur Nedospasov at the Kurchatov Institute in Moscow—the formerly cylindrical plasma had adopted a helical shape and plasma instability had set in. The voltage increased in agreement with theory. In 1962, Kadomstev proposed that the plasma would become turbulent as the magnetic field was raised to a few times the critical level.

My Ph.D. experiment was designed to find out if Kadomstev was correct. Working 1 day a week, nights, and occasional weekends, I assembled the apparatus over the next 18 months in a corner of D-1. The 10-cm-diameter, 2-m-long, positive column, situated in a magnetic field, is shown in Figure 4.1.

The coils, surplus from a previous experiment, as well as computer code to get the correct coil spacing for a uniform field, were provided by Bertie Robson. I used a power supply in an adjacent building, which was surplus from the canceled ICSE project.

A quarter mile north of D-1, the 1MJ of 100 kV capacitors, also purchased for ICSE prior to its cancelation, leaked oil quietly in a hangar. The oily floor beside

Fun in Fusion Research. DOI: http://dx.doi.org/10.1016/B978-0-12-407793-5.00004-2

Figure 4.1 The author looking at the glow discharge.
Source: Photograph reproduced courtesy of UKAEA.

them made a wonderful skid pan for one of our younger engineers—very sober in demeanor, but a nutty auto-racing buff in practice.

The Patrol Force

Up the old runway from D-1 was the control tower, left over from when Culham was a naval air station. It served as the headquarters of the laboratory's patrol force. They served as security and as the site fire brigade; a good bunch of lads but often acted like the Keystone Kops.

Tea and coffee and breaks at 10 a.m. and 3 p.m. were *de rigueur* at Culham. One afternoon, during a break, Roger White and I were sitting with Thundering Sam. He was still testing advanced switches, and he had left his apparatus running while he took a break. Someone rushed in and shouted, "Sam, your apparatus is on fire. I've called the patrol force."

Sam shrugged and muttered, "I'm having my tea."

When it was obvious that Sam wouldn't move, Roger and I ran down to D-1. Sam's HV testing compound was in the middle of a rabbit warren of corridors between metal safety fences. As we arrived, the patrol force, wearing their fireman's hats, pulled up at the corner of the building near my Ph.D. experiment. We watched as they expertly unreeled the fire hose. Four of them formed a line to carry the hose down the outer corridor. It was immediately obvious that they had gone to the wrong corner. There was no access to Sam's compound from where they were running. They disappeared around the end fence. The hose went taut. A few seconds later, it elastically pulled them backward into the corridor, where they fell on their backs. Regrouping, they grabbed the hose and ran back past Roger and me to their fire truck. Roger and I exchanged glances, bobbed our heads like Laurel and Hardy, switched off the power to the compound, found a handheld fire extinguisher, and had the fire out before the firemen returned, thereby saving the HV equipment from a soaking and averting a costly disaster.

Sam turned up later, looking irritated, having made his point that nothing had the right to disturb his tea break. "Now, I suppose I'll have to clear up the mess you made," he said ungraciously.

I'm sure that the patrol force had tea breaks similar to ours. But, in addition, they had their own kitchen where the night patrol could fry up an early breakfast. One night, one of the smoke alarms flashed, indicating a problem in the service duct that ran underneath the central building. They raced to check on it, only to find out that it was a false alarm. An hour later, the alarm came on again. Maybe there was a problem, they concluded, so they tore down the road to check—another false alarm. Back in their headquarters, they started breakfast. Shortly afterward, the alarm came on again. They debated not going, but images of what their chief would say if he found out drove them down the road yet again—still another false alarm. As they returned to their building, they saw smoke and flames erupting from a window. The fat in the pan in which their bacon and sausages had been cooking had caught fire, and the nearby curtains were ablaze. The fire was so fierce that they had to call in the local Abingdon fire brigade to put it out. *Mortification*.

But life for the patrol force had its moments of glory, too. During a time of terrorist activity, all of the government sites were put on alert for problems. Everybody was warned not to leave packages of any kind unattended. One salesman did not pay attention and left his sample briefcase in the main lobby while he went to the restroom. When he returned, the briefcase was gone. Exasperated, he sought help in recovering it.

After a while, the chief appeared carrying a sodden mess with water dripping from numerous holes. "Is this what you were looking for, sir?"

"Yes. What have you done to it?" cried the salesman.

"Standard procedures, sir."

"What procedures?" The salesman was livid.

"As per the recent orders, we placed said object in a bucket of water and, from a safe distance, we shot it."

"In God's name, why?" The salesman was now apoplectic.

"In case it contained explosives." The chief handed the briefcase over. "Next time, please be more careful, sir." The chief clicked his heels, saluted, turned smartly, and strode away. The salesman slumped into a chair, his face in his hands.

In 1964, I married Dace (pronounced *Dat-sa*) Kancbergs, a Latvian-Australian dentist. We bought a run-down thatched cottage in Lower Radley, across a field from the river Thames and a few miles south of Oxford. My wife joined a fellow Australian, David Bradley, in his dental practice in the local town of Abingdon, while I continued to work on Tarantula and my Ph.D. experiment.

Both experiments broke new ground: Tarantula made exquisitely fine shock waves moving perpendicular to the axial magnetic field, behaving in agreement with MHD theory, and I was able to match my nonlinear theory of the plasma behavior in my experiment, an extension of the analyses of other scientists, to my experimental results.

In the first public talk that I had ever given, I told Culham's theory group about the results and analysis of my experiment—an intimidating experience. I was in full stride, having overcome my initial self-consciousness, when the door was flung open and an asthmatic colleague, who had run all the way from D-1, was hanging onto the jamb, trying to speak.

"Your ... wife," he gasped, waving an arm. "Your ... wife ..."

My God, I thought. "What's happened?"

"On phone ... your office."

"Excuse me," I said, ran frantically to D-1, and picked up the phone. "What's the matter?"

"Nothing," Dace replied calmly. "Do you want to play bridge with the Bradleys tonight?"

"That was it?"

"Yes."

"Okay," I said wearily. "I need to get back to my talk."

The theoreticians, none of whom had left, looked at me curiously when I returned.

"Now, as I was saying ..."

5 Culture Shock

By the time I submitted my Ph.D. to London University in 1966, Dace had indicated that she'd prefer to live in a warmer climate. Coincidentally, Roy Bickerton announced that he had accepted an offer from Bill Drummond, a professor in the physics department at the University of Texas in Austin, to spend a sabbatical year there. Roy's objective was to construct a Tarantula-scale experiment that would study shock waves moving at an angle to the magnetic field; similar to those that are found in the bow shock wave of Earth, and other planets like Uranus, where charged particles from the sun hit the planetary magnetic fields. Bertie Robson would accompany him and lead the experimental effort.

I asked Dace if she was interested in going to Texas, where it would be a lot warmer. She questioned whether it would be safe, what with those gun-happy Texans. I assured her that those problems were exaggerated. Shortly after Roy arranged for us to go to Texas, the sniper in the tower at the university went on his rampage. Our marriage survived and armed with my Ph.D., we left England in December 1966. After a brief vacation in Jamaica, we arrived in Austin and rented a house in an old part of town.

It was definitely warmer in Texas. Tony Malein, my former office mate, and his family came the following summer for a sabbatical. They arrived at the Dallas airport and exited the Boeing 727 through the rear door onto the tarmac. It was so hot that Tony decided that the plane's engines, above them, were still running, until he realized that it really was over 100°C even in the shadow cast by the plane.

A year later, Les Woods, a professor at Oxford University, arrived on a sabbatical with his wife and daughters. Les, who was cautious with his money, had found a truly economical flight from London to New York in a de Havilland Comet IV, into which they were packed like Japanese riders on the Tokyo metro. Alcoholic drinks ran out before they reached Scotland, and no food at all remained after they crossed the Western Isles. Worst of all, the water ran out over the Atlantic. Finally, they landed in New York and pulled up near a terminal. Passengers started to exit from the front of the plane. Now, the Comet has the standard tricycle landing gear and, as the front of the plane emptied, the excessive weight of passengers in the rear caused the Comet to tip back on its rear end. After a period of confusion, the captain asked the remaining passengers, including the Woods family, to crawl up the aisle and reestablish the plane's more conventional stance. Les's family was still not speaking to him when they arrived in Austin some 8 hours later.

Another adventure awaited Les during the run-up to the November 1968 elections. On his way up the interstate to give a colloquium at General Dynamics in

Fun in Fusion Research. DOI: http://dx.doi.org/10.1016/B978-0-12-407793-5.00005-4

Fort Worth, he ran out of gas. Soon after he had pulled onto a side road, a man in a pickup truck stopped and came over. "Can I help you, sir?" the elderly man said.

"I've run out of petrol ... uh, gas," said Les.

"There's a station down the road, and I've got a gas can. Come with me, sir," the man replied.

Les climbed into the truck. As they drove 10 miles back down the interstate to the gas station, Les chatted with the kind farmer who had rescued him. "What do you think about the election?" he asked.

"Well, now, my brother fancies that George Wallace," the man replied. "But I don't take to him much."

Thank God for that, thought Les, a true university liberal.

"No, sir," the man said. "To my mind, that George Wallace is a pinko-commie-liberal." He went on to comment on student and faculty rallies in Austin, explaining that he would bring in the state troopers, line'em up, and shoot'em. "Now what do you do, sir?" the farmer asked.

Les, now conscious of the shotgun mounted behind his head, replied casually, "Fluid mechanics stuff, you know ... useful for airplanes. I'm giving a talk at General Dynamics." They finished the ride in silence.

After we had been in Texas for a few months, I found out that Bill Drummond was under the impression that I was also on a sabbatical from Culham Laboratory—a scary discovery for a young couple with a new baby. When I told him that I had resigned from the laboratory, he quickly arranged for me to have a permanent job. *Permanent* didn't really mean anything, though, because I was paid out of research dollars, and if the funding ceased, so would my job—very different from the relative security, at the time, in the UKAEA. Bill also supported my proposal to teach and obtained the position of assistant professor at zero pay for me. I didn't tell him that I needed to do this as a way to learn the physics that I hadn't bothered to learn at Imperial College.

For my first course, I taught sophomore electricity and magnetism. I prayed that my 150 students wouldn't see how scared I was, for this was only the third time in my life I had spoken in public. At the end of the semester, the students rated my performance—not very good. I had been through the British system, which was very different. Beyond the smart remarks, such as "Bloody British accent," and "He write poorly," there were useful tips as to what I needed to change. I consulted my colleagues, and the next semester, taught freshman heat, light, and sound. I received a good rating and also learned a lot about these important topics (Figure 5.1).

Unexpected Diversions

In the fall of 1967, our group hosted the APS's annual plasma physics and fusion research conference. The conference entertainment included a barbecue at a dude ranch outside Austin. Dick Aamodt, who organized the event, had been really

Figure 5.1 Physics department faculty, circa 1969: The chairman, Harold Hansen, is at the right front. Bill Drummond is behind him to the right. My office mate, Gernot Decker, is between Bill and me.

clever. He talked auto dealers in Austin into underwriting a small rodeo as a highlight of the party. The catch: they insisted that at least some of the people at the conference attend a dedication of a new Texas Bluebonnet Trail, a pet project of Lady Bird Johnson. To meet this commitment, Dick arranged for the first two buses heading to the barbecue to make a detour to the dedication.

The information in the program wasn't clear, and the detour planned for the 1:00 and 1:15 buses was known only to the local scientists. Most of the foreigners, not realizing what they were in for, decided to leave early to check out the dude ranch. So a bunch of Europeans, the Japanese attendees, Russians, and a few unfortunate Americans were dropped off by the gym in a high school out in the Hill Country.

Well-dressed elderly women, holding umbrellas to protect them from the never-ending rain, greeted the visitors as they got out of the bus.

"We'all are so glad y'all could come to our dedication."

"What are we doing here?" one of the Europeans asked suspiciously. "I thought we were going to a dude ranch."

"I don't know anything about that," said one of the women, ushering him into a humid, un-air-conditioned gym. Y'all please come in and sign our book."

Another woman indicated a table. "There's Kool-Aid and cookies. Please take some and go to your seats." She pointed to four empty rows in the middle.

The conference attendees joined a bunch of locals dressed in their Sunday best. One of the scientists pulled out a piece of paper and a pencil and started to work on equations.

The lead matron said, "Please stand for our national anthem and 'The Eyes of Texas.' "

The Americans dutifully sang, while the foreigners merely stood, out of respect but looking baffled, and then sat down.

"We welcome all of you to this dedication of a Bluebonnet Trail. It is dear to the heart of our wonderful patron, Lady Bird Johnson," said one of the women who had met them at the door. "We'all are so pleased that we have visitors from many parts of the world. Please, would you stand when I call out y'all's country:

"England. Fray-ance, Germany, Italy." And so on through Europe.

"Japan." The Japanese bowed and made small hissing sounds.

"Arkansas." The foreigners looked more bewildered. Anne Davies, later director of the DOE-OFES, stood and smiled.

"Korea and . . ."

The introduction took a good 15 minutes, and it was followed by speeches, singing by the school choir, more speeches, and more Kool-Aid and cookies until a merciful end came about 2 hours later.

"Y'all come back, y'heah."

"Thank God that's over," muttered someone as everyone got back in the bus. Unfortunately the ordeal wasn't over for one of the groups. The continuous rain had caused the local streams to rise. As the last bus crossed a ford in the road, it stalled.

"Shi-it," said the driver, standing. "Y'all gonna have to take y'all's shoes and socks off, get out of the bus and pu-ush."

"All of us?" the passengers cried in unison.

"Won't move else y'all do *ai-it*."

The water was at least a foot deep and cold. It took nearly 10 minutes to get that battered yellow school bus out of the stream.

The attendees who'd avoided the dedication arrived at the ranch starting at 3:30. They were already liquored up and happy, despite the fact that the rain had led to the cancellation of both the rodeo and horseback riding. They cheered when the first of the dedication buses unloaded a group of grumpy campers, who went immediately for the booze. As for Dick Aamodt, disappointed that his efforts had gone awry, he grabbed a bottle and retired to the base of a large tree.

A day later, I heard about another curious event. After the barbecue, Bill Drummond drove three of the visitors back to Austin. "I've got to make this phone call," Bill said as they entered the town of Wimberly. He didn't explain why, but his colleagues were used to his hectic, right-now way of doing business. In the town, he couldn't find a pay phone, and nothing was open that late. (This was before cell phones.) As they continued toward Austin, they passed a house with all its lights on, including the ones by the front door.

"I'll see if they'll let me use their phone," Bill said, stopping the car. He rushed to the front door. An elderly woman appeared and, as far as the passengers could see, before Bill could say anything, ushered him in. A moment later, he reappeared and signaled for his guests to join him.

The plump lady who met them at the door was wearing a burnt-orange cocktail dress, and her blue hair was puffed out in a 1940s style.

"I'm so glad y'all were able to make it." She beamed. "I was worried y'all had forgotten the date."

Seeing someone about to say that they weren't who she was expecting, Bill shook his head and said, "We're sure glad to be here, ma'am."

She ushered them into the parlor. The tables were bulging with all kinds of food, a large bowl contained red punch with ice, and nearby were assorted soft drinks and coffee. Party decorations festooned the walls. It looked like they had gate-crashed a party . . . except that, in the time they stayed, nobody else came.

Out of courtesy, they stayed for about a half-hour, eating, drinking, and making awkward conversation. Then they made their apologies and left.

Back on the road, they talked about what had happened. "I thought we were in an episode of *The Twilight Zone*," someone said. Another person wondered if they'd been in a fatal wreck and this party was the beginning of the afterlife. It was a relief to see the Austin city lights.

Bill called around to his many contacts and learned the explanation for the strange event. The poor lady's life had revolved around her husband. After he died, giving parties was her therapy; but, sadly, Bill and his friends were the only ones who had ever turned up.

The Oblique Shock-Wave Experiment

Trying to build an experiment at the university was an eye-opening experience for all of us. We had been used to the support structure at Culham, with its well-equipped stores, large workshop, glass-blowing facilities, welders, and electronics shop. The physics department at the University of Texas had a modest-scale workshop with experienced machinists, but most materials had to be brought in from Houston. Fortunately, we had a general factotum, John "Dutch" Goebel, and his assistant, Richard, a skinny young man with such incredibly strong hands that we soon coined the phrase, "What Richard has tightened up, let no man undo."

With Bertie's inventive skills, hard work, perseverance, and complete drawings for a Culham-designed HV switch, we managed to have the collisionless, oblique shock-wave experiment operating within the year. The shock tube layout was similar to Tarantula, with a magnetic field along the axis. The main difference from Tarantula was the segmented copper band (shock loop) around the mid-plane of the glass vacuum vessel that drove the shock waves.

Jim Paul's group had discovered a foot appearing in front of the shock on Tarantula. This phenomenon was also seen on our experiment. Perry Phillips, Bertie's graduate student, built a probe to collect ions coming out of the shock. With the results in hand, Bertie and Al Macmahon explained the phenomenon as being due to ions, accelerated within the shock front, orbiting around the magnetic field. Decades later, the same foot was discovered by a probe traversing the bow shock of the planet Uranus. We also measured whistler waves coming off the oblique part of the front—a name given to the sound that such ionospheric waves make in radio signals.

Another significant visitor was the distinguished British physicist G. I. Taylor, of Rayleigh-Taylor instability fame. At age 81, he gave a lecture on his recent work on the effects of an electric field on water droplets—experiment and theory— that was a model of how to do research.

Two States Separated by a Common Language

Texans are charming, but they generally have a deaf ear when it comes to different accents and indicate it by saying, "Sir?" with a puzzled expression. We also got used to hearing "y'all," as in "Y'all curm back y'hear." It was Bertie who discovered, on leaving a store with his family, that there existed a plural form. The salesman called out after them, "Yousel'all come back."

Bertie thought about it for a moment before replying, "Wesel'all sure will."

"Sir?"

Geography is also an uncertain subject. Our landlady explained that she would be visiting her daughter in Frankfurt, Germany, and would travel via Paris and Houston. A circuitous route, we thought. Then we realized that the Paris she meant was not the one in France—it was a city in Texas.

Another English expatriate was talking to an elderly Texan woman who commented, "You have a strange accent."

Our friend explained, "I came from England 6 months ago."

"My, you've picked up the language quickly," the lady exclaimed.

One among our many British colleagues was very proud that his son had retained a British accent. I was surprised, one day, when I found him sitting at his desk looking unhappy.

"What's the problem?" I said.

"They all seem charming enough, these Texans. But ..."

"What happened?"

"They're devious ... devious, you see."

"What?"

"I dropped Charles off at his school today, as I usually do, and he said, 'Goodbye, Daddy.'"

"So?"

"Then he saw this little friend across the road. He waved, 'Hi thair, Cindy Lou,' he said. Damn it, the kid's bilingual! They got to him! It's not acceptable."

I shouldn't have laughed. I should have realized that children are very sensitive to appearing different from their peers. Some years later, when we returned to rural Berkshire, Dace overheard our son Jason practicing his new English outside the kitchen window.

"Ah murst go dow-un the la-ain with ma fra-inds, and look at the trurck," he said. "No, no. I murst go doon the lane with moi frainds and see the lorry. No. No ..."

Continuing with the subject of two countries divided by a common language, during my stay in Texas, Bertie and I was nominated to buy our group's secretary, Loretta, a birthday present. Everybody in the group contributed, and we decided to get what people in England and some parts of America call a *brooch*. Our colleagues warned us that a more common name in Texas was a *pin*. We headed downtown to the main drag by the Capitol in Austin and started with a jewelry store at the bottom of the street nearest the Colorado River.

"I would like to buy a pin for our secretary," Bertie said to the middle-aged salesman.

"Sir, would that be a Parker pain or a Schaeffer pain?"

"No," he said. "I mean a lady's pain. I would call it a brooch."

"Sir?"

"To go on a lady's dress." Bertie indicated the whereabouts on his lapel.

"You mean a pai-ain?" This time, he appeared to be using more syllables.

"Yes."

He showed us his selection. They were too cheap-looking, though, so we left.

In the next store, when the salesman came up, I said. "I would like to bah a lai-aidy's pai-ain." I indicated my lapel.

"We sell Parker pai-ins, sir."

"No, a pai-ain that goes on a dress. I would call it a brooch."

He looked puzzled for a moment before grasping my meaning. "Y'all want a pai-ai-ain."

The selection was slightly better than the previous shop, but they had nothing we liked.

As we proceeded up the street, this bizarre procedure was repeated in two more jewelers. We fought off their attempts to sell us pens, with our accents becoming more and more exaggerated as we walked to the last store. At the final jewelers, when the salesman approached, I swaggered up to him, and said. "Sir, ah would lahk to purch-ai-ais a pai-ai-ai-ain for our secretary, Lor-ret-tah."

He looked at me disdainfully. "Sir, would y'all be looking for a brooch?"

We gaped at him. "Yes."

"We have a fine collection, sir."

They did, and we found the perfect present and bought it.

6 Talks Can Surprise Us

The most hilarious lecture I ever heard was given by George Yevick at a meeting in Los Alamos in 1968. George was a professor at the Stevens Institute in Hoboken. The way he gave a talk, George should have been working the Borscht Belt circuit. With his determined delivery, he sounded like a tailor trying to sell you a second pair of trousers to go with the three-piece suit you were buying. With apologies to George, if I don't have it quite right, I will do my best to recapture the memorable and informative talk about his efforts to build a delicate, optical scientific instrument.

"So, I want to build this interferometer. Not your run-of-the-mill instrument, you understand: an interferometer to lead the world." George, an ebullient, small man dressed in a neat dark suit, looked around to make sure we understood.

"A cousin found low-cost naval optics. Our navy, you can be sure."

George put his hands in his pockets and they came out with some small change. "But, and I tell you the truth, I had little money left ... only 122 dollars to build a stable operating platform for this magnificent device, my 'Immaculate Interferometer.' "

George paused again for effect. "What to do? What to do? My laboratory is in a building next to a main highway. Trucks and buses all the time shaking the building, rattling my laboratory on the second floor. I should live in an earthquake zone. It would be quieter." He wrung his hands to show how desperate was his predicament.

"What did I do, you ask?" He looked at us, as if we would know the answer.

Seeing only interest and amusement at his tale, he continued. "I went to five neighborhood Woolcos and I bought all their large rubber bands. I strung them together to make two harnesses." He reached into his bag and produced a bunch of bands, which he stretched to show their strength and elasticity. We clapped to show our appreciation.

"But that cost me 26 dollars and 23 cents, and with other incidentals and my bus fares, I had only 95 dollars and 38 cents left." He raised his hands in despair. "I still had to find a solid table to support the spectrometer. I needed flat and smooth granite or marble. Who can afford it?" He grimaced to show the magnitude of his predicament.

"What did you do?" a number of us chanted.

"Well might you ask. Fortunately, I remembered my friend, Mordy, the undertaker.

'What can I do, Mordy? You must help me,' I pleaded with him. Mordy scratched his head, and then came up with the solution. 'Unused gravestone,' he

Fun in Fusion Research. DOI: http://dx.doi.org/10.1016/B978-0-12-407793-5.00006-6

answered. 'Why would there be an unused gravestone?' I asked. 'People think they're going to die. Prepare to have a nice carving done. Sometimes it doesn't work out,' Mordy raised his hands in despair. 'They live! Later, they don't like the inscription. It doesn't fit their new image of themselves. So ... unused gravestone.' "

"Parenthetically, Mordy has some great deals if you're ever in need." George looked around, as if ready to take orders.

When we did not respond, he continued. " 'How do you know it hasn't been used?' I asked. 'Experience, George. If people thought I was selling used ones, it would not be good for business.' "

"Now, I will show you the completed masterpiece." A slide appeared. In it we could see the Immaculate Interferometer, mounted on a marble gravestone—unused—and hanging from a bunch of rubber bands.

George, looking triumphant, pointed at the picture. "And I still had nine dollars and five cents left."

We cheered.

Most talks are not as interesting and entertaining as Professor Yevick's. A colleague, Ram Varma, told me this story about an unfortunate colloquium given some years earlier at the University of California San Diego (UCSD). A university scientist from an eastern college had discovered what he believed to be a new effect in the positive column of a glow discharge, an experiment similar to my Ph.D. setup. He had spent his summer touring the country, getting a warm reception when he related his findings and his explanation of the phenomenon. He was doing fine until he reached UCSD, where he encountered a distinguished German scientist, Professor Ecker—one of the fathers of the field of gas discharge physics.

The speaker had spoken for only a few minutes, giving the gist of his discovery, when Ecker stood up. "Young man," he said, "the problem with researchers of your generation is that you are unacquainted with the literature. If you had chosen to read my article, in the 1927 edition of *Zeitschrift fur Physik*, you would have learned that I observed this phenomenon and explained it over 30 years ago."

Professor Ecker sat down.

The speaker's voice faltered, seeing his career in ruins; then he wobbled and then fainted flat out on the stage.

The colloquium organizer rushed over to check on the man, who was showing signs of recovery. The audience sat in stunned silence. The organizer signaled for help, and he and Ram assisted the speaker out of the lecture theater. The man did not continue his tour.

Fortunately, there were some real advances made in the 1960s. In 1968, scientists at the Kurchatov Institute in Moscow had a breakthrough with their T-3 tokamak. At the conference in Novosibirsk, Siberia, the director, Lev Artsimovich, announced that they had achieved an electron temperature of 10 million °C (1 keV). The news was met with skepticism by many of the world's fusion scientists because the institute's methods of measurement were open to interpretation. Also, the scientists had not used the latest diagnostic technique, which employed the scattering of laser light from the electrons to determine the temperature.

Because the scattered light is less than 1 million millionths of the input beam, such measurements only became possible, as I mentioned before, with the development of lasers that could produce a joule or two of light. Most notable, at the time, was a ruby laser that generated red light. Even with these lasers, each detected signal contained only a few hundred photons.

The leaders of the British delegation from Culham Laboratory negotiated to send a team of experts to Moscow with scattering equipment to make an independent measurement. The data collected later that year confirmed the Russians' success, and soon tokamak experiments were being constructed worldwide.

There was a similar issue with inadequate measurement of the electron temperature on our shock-wave experiment. We decided to install a Thomson scattering system using a novel ruby laser that had been developed at the Naval Research Laboratory (NRL) in Washington, D.C. I was sent there to learn about the laser. Armed with the information I obtained, Bertie and I built the laser and the accompanying transmission and collection optics to allow us to scatter the laser light off the electrons in the shock front.

Our measurements of the signal scattered at $90°C$ to the beam confirmed our previous estimates of the electron temperature around $30\,eV$. Inspired by this success, we decided to measure the light scattered at small angles to the beam. Forward scattering is a more difficult measurement because the laser light itself can swamp the scattered signal. In these measurements, the light may be scattered off density oscillations caused by instabilities—the fine-scale instabilities that we expected would be driven in our collisionless shock front. The shock thickness of just a few millimeters was less than the particle collision lengths. We employed a new graduate student named Milt Machalek to do the work for his Ph.D.

After a year setting up the experiment, and following the birth of our second son, Nicholas, Dace and I decided to return to England. She would return to David Bradley's dental practice in Abingdon, and I would go back to the Culham Laboratory. Although we were sorry to leave Austin, where we had many friends, it would have been difficult for Dace to qualify to practice dentistry in Texas. The nearest dental school was Galveston, and we had two young children.

Unfortunately, I was the only person working on Thomson scattering in the physics department. Concerned about support for Milt now that I would be leaving, I gave a graduate course on scattering and wrote a comprehensive set of notes for Milt. During this time, a literary agent visited and roamed the halls, asking if anyone wanted to write a technical book. Off the cuff, I said "Yes, *Scattering of Radiation from Plasmas.*" He asked for a one-page description of the contents, and in short order, he obtained a contract for me with Academic Press.

By the time I left Texas, Bertie was investigating how to replace the shock-wave experiment with a tokamak, and Clifford Gardner, a brilliant mathematician, had arrived to spend the summer in Bill Drummond's group. Clifford had a diffident personality and a reluctance to display his ability. One day, I went to him for advice on a mathematical problem. "Can you help me, Clifford?" I asked. "I'm sure it's possible to solve this problem analytically, but I'm stuck."

Clifford's head sank, tortoiselike, into his collar. "I don't really know anything about that," he said apologetically.

"Isn't this what Gardner's theorem is about?" I asked.

"I suppose so."

"Aren't you that Gardner?"

"I suppose so," Clifford replied cautiously. Then, appearing to be resigned to having to demonstrate his abilities, he patiently showed me how to deal with my problem.

Cakes and Ale

Near the end of my stay in Texas, Bill Drummond managed to persuade the DOE to support building a tokamak at the university. Initially known as El Toro Grande but referred to by some others as a "Load of Bull," it was finally called the Texas Tokamak (TEXT). Bill announced that he would have a party at his house to celebrate his success, and he wanted to have a toroidal cake in the image of the tokamak. Ken Gentle volunteered to make the cake. He and Dutch Goebel made cardboard models of a set of toroidal coils. I heard a rumor that Ken talked Bill out of his additional fantasy of mounting this edifice on a floating platform in his pool and launching it upward with rockets.

On a very hot Saturday afternoon, Bill's entire group, supplemented by various other people from the university, assembled on Bill's patio. Ken's cake was magnificent—*see the cover picture*. He had mounted an array of 24 sponge cakes, each 10 in in diameter, on a plexiglass ring inside the toroidal coil set. Ken had covered this torus, nearly the size of that planned for the TEXT vacuum vessel, with thick, burnt-orange frosting, the color of the University of Texas. He had filled the cracks with additional icing (Figure 6.1).

The huge cake sat on a table at the edge of Bill's prize zoysia-grass lawn that gently sloped down to the creek. After some hours of eating barbecue and drinking, the attendees turned their attention to the cake. By now, the icing had started to melt, and the cake was sagging toward the table. As Ken cut off the first piece to give to Bill, a section of cake fell to the ground and rolled onto the zoysia. The audience cheered, and Ken acted rapidly to cut off enough slices from the remaining cake to feed everyone. As we watched in drunken amusement, the whole mass became unstable, and large, burnt-orange chunks fell off the table and wandered down the lawn.

"Don't worry, the birds'll eat it," someone said, then added "Wait, look, the icing's burning the grass!"

"The apricot jelly was too acidic, I guess," said Ken, studying the leprous-looking, orange-colored streaks on the lawn.

"Don't worry," said Bill. "I'll get it replaced. And next time, we'll put the cake further back on the patio."

"What next time?" Ken asked.

Figure 6.1 Ken Gentle's cake.

"When we get the super upgrade I'm thinking about," replied Bill.

Ken Gentle seemed to have an aversion to what he felt were ill-considered comments. Consequently, when Dick Aamodt announced that he drank only Gordon's gin and could tell what brand of gin was in his martini simply from the taste, Ken decided to challenge his assertion. He invited four couples to his house for dinner, including the Aamodts and Dace and me, but he did not tell us the reason for the invitation. That became clear after we were seated in his living room.

Ken handed each of us a pencil and a piece of paper with our name on it. Each paper was divided into sections numbered one through four. "We will start with four martinis," he said, grinning. "They're made identically, except three have Gordon's, Booth's, and Bombay gin, respectively, and the fourth is a repeat of one of them."

"Why are we doing this?" I asked.

"Dick knows," Ken replied. "What I'd like you to do is write down what you think of the martini, which brand of gin was used, and which one was repeated."

He went to his kitchen and returned with eight martinis, each with a numbered label, and handed them out.

Over the next hour, we sipped our drinks and made notes. Interestingly, the martinis did taste different; specifically, the cheapest one was harsher. Mind you, it became hard to concentrate on our task after a bit. Finally, Ken collected our responses and disappeared.

After a while he returned and announced his findings. "A few of you correctly guessed which gin was used. Only two of you identified which gin was repeated. You agreed quite well on which one you liked and which you didn't."

"How did I do?" Dick asked.

"You didn't get any right." Ken's straight face could not hide his amusement. "Let's go in for dinner. I've got a really good wine."

Six months later, I was sitting with Ken and Per Nielsen at lunch when Per informed us that Pearl beer was by far the best he'd come across in the States, and he could tell it blindfolded. I knew from Ken's expression what was going to happen. A few weeks later, Ken invited 10 couples to his house. He had developed a more extensive questionnaire, and fed us eight types of beer with two repeated, including Pearl, Lone Star, Budweiser, Miller, Pabst, Old Milwaukee, and Schlitz. Our Texan beer, Pearl, did rank highly on taste. Per didn't identify it, and a few people worked out which two beers were repeated. Interestingly, the cheapest beer, Old Milwaukee, also ranked highly.

Later, Per and I discussed getting someone to make an ill-advised comment to Ken about fine French wines, but we did not follow through on our proposed scam. Incidentally, Per is in the middle of the photograph on the cover with his back to the viewer and facing his blonde wife, Annie.

In 1970, before returning to England, I attended an international conference in Rome. The early years of fusion research were marked by many inspired, but slightly nutty, experiments. One that I heard about during the meeting was at the Culham Laboratory. It involved a concoction of copper coils suspended in a cylindrical metal vacuum chamber. The coils carried large currents of electricity to produce a subtle arrangement of magnetic fields to contain plasma. The facility was appropriately named CLIMAX.

Harry Megaw, an Irish physicist from Culham, related what happened to CLIMAX when the engineers switched on the experiment.

"Unfortunately, our engineers had not calculated the forces correctly," he said. "A sound, reminiscent of the percussion sections of a dozen symphony orchestras, filled the experimental hall, as all the little coils wrapped around each other." He showed a picture of a tangled mess of copper conductors.

We all smiled. There is something about the lilt of an Irish accent that makes it hard to take descriptions of disaster seriously.

7 Culham Again

Neutral Beams and CLEO

When I applied to return to Culham, I anticipated that they would want me to work on the scattering diagnostic on which I had gained expertise in Texas. However, the laboratory had plenty of experts, including the ones who had made the crucial measurement on the Russian T-3 tokamak, so I was assigned to work on the development of intense beams of neutral particles that would be used to heat plasmas. In such systems, a plasma source is used to create ions of the chosen species (e.g., hydrogen or helium). The ions are accelerated by a set of electrodes, and then a fraction is converted to neutral atoms by passing them through a gas cell. The excess ions are swept out of the beam by a magnetic field, and the neutrals are directed down an evacuated tube to heat plasma (in our case, initially in a tokamak and subsequently in a stellarator).

Owing to the existence of historical fiefdoms, there were three groups at Culham working in this area. Each team was developing hardware for application in its division, a ridiculous situation that was known worldwide. Bob Pyle, who ran the beam research at Lawrence Berkeley National Laboratory in California, alluded to this situation at an international conference. "I find it strange that half of the world's neutral beam development groups are at Culham," he said to an amused audience.

Nevertheless, it worked out well for me, for I was assigned to work in the group headed by Bert Cole, with the goal of developing a 20 kV, 1 A, neutralized hydrogen beam for injection into the CLEO experiments. Because Bert's interest lay in another research area, I in effect led the development of the beam system. We based our program on beam systems developed by Bill Morgan's group at the Oak Ridge National Laboratory—they had kindly provided us with drawings of a plasma source and accelerating electrodes. We combined their plasma source with electrodes based upon designs developed at Culham. Eventually, we produced an 8.5 A beam at 30 kV and performed the world's first tokamak neutral beam heating test on the CLEO tokamak in early 1972. To show that the injected particles were behaving as expected, we analyzed the spectrum of energetic hydrogen ions leaving the plasma by charge exchange with neutral hydrogen. We were surprised to find that there were particles with energies greater than the injected particle energy. Geoff Cordey calculated the expected spectrum, including the fact that there were electrons moving faster than the injected particles. He showed that the high-energy tail was caused by these fast electrons, and that the overall measured spectrum was

Fun in Fusion Research. DOI: http://dx.doi.org/10.1016/B978-0-12-407793-5.00007-8

in agreement with the theoretical model. Later, when we increased the heating power, we raised the plasma temperature. The early success of this heating system on CLEO and the tokamaks, ATC at Princeton and ORMAK at Oak Ridge led to a massive effort to make more powerful neutral beam systems. Our systems, which were in the range of tens of kilowatts, were the forerunners of today's multimegawatt systems.

A Weird Occurrence

To compensate for reductions in its fusion program, Culham had developed programs in areas that could take advantage of its technology expertise. As I said earlier, the lab had a number of HV, high-current systems. One of the researchers had the bright idea that one of these systems could be used to simulate a lightning strike to an aircraft. When a plane is hit, the current runs through the skin of the wings, tail, or fuselage. If that's all that happens, there's no problem. But if there's a bad electrical connection, you get an arc, and if it's near wiring or fuel tanks, that's a big problem.

He received funding from British organizations, and soon he was testing a Concorde nose cone and the fuselage of a jet fighter. He then made a proposal to the European air safety organization in Geneva to find ways to improve electrical connections in the skin. One of their experts visited. The expert was not impressed with our work, saying that Culham's puny apparatus was no match for real lightning. He left at lunchtime, muttering that it had been a waste of his time.

To everyone's surprise, he called later in the afternoon and said that he would be recommending funding for the proposal.

"What changed your mind?" asked my colleague.

"As we were coming in to land at Geneva airport, my plane was struck by lightning ... twice," the man replied. "I'm not going to argue with God."

8 JET: Larger and Larger

In 1971, buoyed by the success of the world's tokamak programs, the European Union's fusion program initiated a study of options for a major next-step tokamak— the Joint European Torus (JET). In 1972, the so-called Luc Committee, with members from the laboratories partnering in the European effort, recommended that a tokamak with 3 MA of plasma current should be designed. At this time, the world's largest operating tokamaks had far less than 1 MA. The reason for the choice was simple: At this current, the majority of the alpha particles produced by the fusion of deuterium (D) and tritium (T) would be contained in the plasma and slow down the heating of it—the goal of a self-sustaining fusion system. So, the proposed device would have not only a huge current but also operation with D−T—this would be a bold step.

I attended the meeting at Culham in the summer of 1973 at which three designs were discussed: A design from the Italian Frascati laboratory, another design, led by Alan Gibson of Culham, that used superconducting coils, and last, a very innovative design from the French laboratory at Fontenay aux Roses, developed under the leadership of Paul Henri Rebut. Rebut's design had a noncircular plasma cross section, when nearly every other tokamak's had been circular. It had a low aspect ratio (i.e., the ratio of the major radius to the minor radius of the torus) of 2.4, while most tokamaks were in the 3−6 range. A bicycle tire has a large aspect ratio, an automobile tire an intermediate ratio, and a doughnut a very low aspect ratio (Figure 8.1).

The follow-up to the meeting was to choose the option to go on to a fully engineered design and, ultimately, construction. The international design team would be housed at Culham; the site for construction would be decided later. Prior to the meeting, Alan Gibson had recruited me from the neutral beam development group to work with him on JET issues. This was a great move for me because Alan was one of the most practically inventive and bright scientists at Culham. I asked him whether I would be working on neutral beams or, possibly, scattering measurements. At this time, I was close to completing my book on scattering. He replied, "No. I'd like you to work on cost analysis of tokamaks. Check whether Paul Henri's right about the aspect ratio."

Alan had developed the first cost analysis code to include physics goals, component costs, engineering constraints, and site constraints to optimize the DITE tokamak, which was then under construction in D-1. "What exactly do I have to do?" I asked.

Fun in Fusion Research. DOI: http://dx.doi.org/10.1016/B978-0-12-407793-5.00008-X

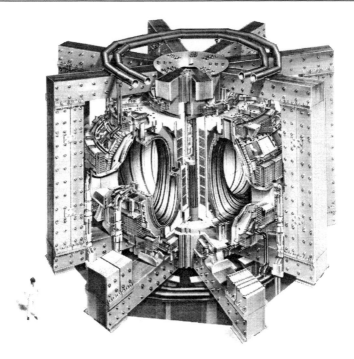

Figure 8.1 Artist's drawing of an early JET design.
Source: Courtesy of the JET organization.

"Upgrade my DITE code to include options for noncircular plasmas and noncircular toroidal coils, and add more physics models," Alan replied. "We need to analyze the cost-effectiveness of the proposed three options for JET."

With help from Hugh Crawley on engineering issues, and with advice from
Alan and other physicists, I upgraded the code. It is important to understand that at
the time, there were no good explanations of tokamak plasma heat and particle confinement. Measured losses were clearly much higher than predicted by the so-
called neoclassical theory of transport due to collisions. The theories of various
microinstabilities indicated that many of them might enhance transport, but there
were no measurements to show that they were the cause of the heat and particle
losses. Limits on performance (e.g., the ratio of plasma pressure to magnetic pressure: beta) had been predicted by MHD theory, but no tokamak had been heated to
significant beta to do a test. Nevertheless, it was possible to show that all of these
potential transport mechanisms and limits could be written in terms of a function of
plasma current times the aspect ratio to some power, i.e., $I(R/a)^s$, where s ranged
from 0 to 2.

I was able to show quickly that Paul Henri's design was well optimized regard-
less of which of these scaling laws applied. In fact, Paul Henri had shown on one
page how, for a fixed plasma current and safety factor, the minimum cost would be

at an aspect ratio of around 2.5. The need to have room for a transformer coil in the bore of the tokamak, a large enough space to drive and sustain the plasma current, caused the cost to rise as the aspect ratio decreased. As the aspect ratio increased, it became more expensive to achieve the desired plasma current because the experiment had to be larger to reach the 3 MA current requirement. We readily accepted Paul Henri's design.

The JET design team was well established by the end of 1973, with Paul Rebut as director, Alan Gibson as head of physics, Enso Bertollini (Italy) as head of electrical systems, Michel Huguet (France) as head of mechanical systems, Dieter Eckart (Germany) as head of vacuum systems, etc., and Jean Pierre Poffe (Belgium) as head of administration. By the end of 1974, there were 45 of us—a mixture of scientists, engineers, draftsmen, and secretaries from some nine European countries.

The work was stimulating, and I enjoyed being involved with this team of bright people—a group better than what could have been established by any single European laboratory. My role in Alan's group evolved to deal with the physics of heating systems and diagnostics. These systems would be developed by the European laboratories, and we held workshops to get feedback into what the experts felt that JET should incorporate.

The area of heating was a problem because some people, including Paul Rebut and Bas Pease, director of Culham, believed that the heating by the plasma current would be sufficient to raise the plasma temperature to a required fusion level of 10 kV or more. Other people, notably Dr. Dei-cas, a scientist from Fontenay aux Roses, argued that we might need up to 60 MW of power in addition to the current heating. Dei-cas got it right! To deal with such a divergent set of opinions, we calculated how much power could be introduced through the large ports on the midplane of the tokamak for each of the potential auxiliary heating systems. At the time, these included neutral beams and heating with a range of frequencies from a megahertz to many gigahertz. We calculated that if we were allocated eight ports, it would be possible, by one means or another, to deliver the upper level of power.

Unfortunately, the JET Council, probably persuaded by financial considerations, would permit us to have only 5 MW initially, rising to 10 MW in the first phase of construction, with the possibility of 25 MW in a second phase. This restriction was finessed by redefining the power as that being delivered to the plasma center, so 10 MW became, in effect, 16 MW, and 25 MW became 40 MW of power delivered through the plasma boundary.

In another area, an interesting development had been announced at a 1972 meeting in Grenoble, illustrating how much the Russian scientific capabilities were underestimated. At the time, power sources at 20 GHz and above, in the range of hundreds of kilowatts, were only a dream to most scientists. Then, Dr. Alikaev gave a talk and mentioned that his group had put 100 kW of power at 35 GHz into a tokamak.

After a brief silence, the audience reacted with disbelief.

"He means 35 megahertz," somebody shouted.

Alikaev looked at him and said firmly, "Is gigahertz."

"There's no such thing." The questioner sounded incredulous. "What did you use?"

"Is gyroton?"

"What's that?" someone else asked.

Alikaev went to the blackboard and sketched a device in which a beam of electrons was passed down an evacuated tube with a strong axial magnetic field. High-frequency power was extracted at multiples of the electron cyclotron frequency. In fact, the system, invented years earlier by an Australian scientist, Dr. Twiss, had received attention in the West, but the research programs had not been successful. Stimulated by military interest in a number of applications, including having a source of power at frequencies able to penetrate clouds, Russian scientists at the Gorki Research Institute had been successful.

JET Travels

One of the perks of working on JET was the extensive foreign travel required to keep in touch with the partner laboratories: particularly, Fontenay aux Rose in Paris and Grenoble; Garching, near Munich and Julich; and Frascati, in the hills south of Rome. By the end of 1974, other countries were working on JET-scale tokamak designs: JT-60 in Japan; the Tokamak Fusion Test Reactor (TFTR) at Princeton, New Jersey; and T-20 in the Soviet Union. The International Atomic Energy Agency (IAEA) initiated the large tokamak meetings to facilitate the exchange of information between the four design teams.

Within the European system, I was assigned to be the JET representative to the Euratom Heating and Injection Advisory Committee. Other members came from the European laboratories engaged in this area; the ones mentioned above and also Culham, Jutphaas in the Netherlands, and Riso in Denmark. The significance of the Euratom involvement was that they generally supplied 20% of the budgets of fusion laboratories. However, if a program was determined to have particular merit, it could receive 45% support (an unusual exception was JET, which as an Euratom project received 90% support).

As a whole, the advisory committee meetings were congenial. That continued until we met for 2.5 days to consider an unprecedented request. Up to this point, we had only been asked to comment on scientific questions. This time, we were asked to recommend what percentages of the Euratom budget in this area should be devoted to each of the candidate heating and fueling systems. Our chairman, Chris Watson from Culham, listened with increasing concern as our 16 members argued for a day and a half, each mainly defending his own territory. JET needed development of nearly all of the options. As the JET representative, I had no axe to grind; neither did Chris, a theoretician.

On the afternoon of the second day, faced with no obvious answer to the question that we had been asked, Chris called the meeting to order and said, "Gentlemen, I have to present our findings to the Euratom Fusion Management

Committee the day after tomorrow. We have less than a day to agree on our recommendations and write a report. I have been unable to discern the consensus position from our discussions; therefore, I will hand each of you a piece of paper, numbered 1 to 16. Please write down what percentage of the budget you believe should be allocated to each of the following areas: positive ion beams, negative ion beams, ion cyclotron heating, lower hybrid heating, electron cyclotron heating, pellet fueling, and cluster fueling."

A vigorous debate followed his request, with disagreement being stated in numerous languages (we're not children, why not a lottery? and so on). Chris listened patiently before pushing the pile of paper into the center of the table. "I suggest that each of you write in his views," he said firmly, his tone indicating his irritation with the response. "I will work with what I receive. If nobody responds, I will have to use my own judgment."

One by one, the committee members took a piece of paper and filled it in. I collected them and helped Chris add up the responses. The results were clear. With the exception of one member who wrote, "90% on lower hybrid and I don't care what you do with the rest," the other 15 members were in close agreement. They recommended allocating 40% to beams (mainly positive ions), 40% to heating at the various frequencies, mainly split between ion cyclotron and lower hybrid heating (gyrotron development was still funded at a low level in Europe), and 20% to fueling. The members congratulated Chris on his approach, and many admitted that they had misjudged its effectiveness.

This was my first exposure to the so-called Delphi approach. I used it later with similar success in committee work where the debate of an issue had not led to a clear conclusion.

On another notable occasion, I had to attend a February meeting on heating systems at Garching in Germany, finishing on a Friday afternoon, but then go back to England, only to return to the continent on Sunday for a meeting at the Grenoble laboratory in France. I discussed the travel options with a Culham colleague, Ernie Thompson, who faced a similar schedule.

"If they let us stay, we could go skiing," Ernie said.

"Where?" I asked.

"La Plagne," Ernie replied. "It's one of these new resorts ... somewhere between Geneva and Grenoble. I went there with Jean Jaquinot. He and his brother have a place there. Fantastic skiing."

Agreeing that it made no sense to return to England for a day and a half, we planned to ski La Plagne on Saturday and ski Chamrousse, near Grenoble, on Sunday.

On the Friday night after the first meeting, we flew to Geneva. I got a rental car and we stayed for the night. The next morning, to my surprise, the only La Plagne that we could find on our map was not in the High Alps, but in the Massif Chartreuse to the west. "Are you sure about this?" I asked Ernie.

"Don't worry," he said. "I'll recognize it when we get near. It's at the end of a small road off the main road in a valley."

After breakfast, we left early on the road from Geneva to Grenoble. We were encouraged by large signs with healthy-looking skiers that read "Ski La Plagne."

We also saw numerous cars with ski racks. But things changed when we turned right, up into the Massif Chartreuse, rather than left, into the High Alps.

As I drove along the small road, surrounded by snow-covered fields leading up to craggy mountains, I questioned Ernie. "We haven't seen any signs or skiers in miles. Are you sure this is the right place?"

"Don't worry," Ernie said. "It was like this when I went with Jean."

Finally, we spotted a tiny wooden sign, with "La Plagne" handwritten on it, that pointed left to a tiny road that led up a steep slope. Again I asked, "Are you sure this is it? Look at the road. Hardly any cars have been up here."

"It was like this. Carry on. We might as well see where it goes."

Climbing farther, we passed through a small village to reach a track heading up the mountain to what looked like a farm. As we pulled through a gate, an old man emerged from a house and came toward us.

Ernie opened his window and said, "Ici, La Plagne?"

"Oui," the old man replied.

Ernie turned to me. "How do you say, 'Can we ski here?'" he asked.

I leaned across and said, "Est ce que nous pouvons faire du ski ici?"

"Vous voudriez faire du ski . . . ici?" the old man asked. "Dans ma ferme?"

"I think he asked if we want to ski in his farm," Ernie said to me. He turned to the window and said, "Oui."

"Vous êtes d'Angleterre?"

He wants to know if we're from England," Ernie said, adding, "Oui."

"D'Angleterre pour faire du ski La Plagne." The old man's face lit up, and he smiled, displaying an array of steel teeth. "Ici, n'est pas Ski La Plagne, c'est cent kilometre la bas." He pointed east, in the direction of the High Alps.

"Bloody hell, Ernie!" I yelled. "We're 100 km off—"

"Merci, monsieur," Ernie said, as he rolled up his window. "Let's get out of here. There's got to be some ski place down the road."

As I drove down the mountain, I looked in the rearview mirror. The old man was hightailing it after us. No doubt, to tell all his friends in the village about the crazy Englishmen who had driven hundreds of miles to ski in his farm.

The next day started off better. Ernie and I found the real Chamrousse, a ski resort that had hosted the Winter Olympics a few years earlier. We headed to a chairlift, and Ernie got on. I managed to follow a few chairs behind. Luckily, there wasn't anyone else waiting to get on. I pulled down the bar and rested my skis on it, as I'd been taught.

I watched Ernie ski off at the end of the ride and then go down a path cut into the side of the hill. The lift operator was standing by the small hut where the lift turned around, a lighted Gauloise cigarette hanging out of his mouth.

I hated skiing off a chairlift and thought of all the things I could do wrong as I lifted the bar. I hadn't thought of the one that actually happened. The chair was made of curved metal tubes, and the one supporting the right side of my seat had come loose. Its end projected to where my right leg hung down. At the landing zone, I launched myself forward. My body moved, but I didn't leave the chair behind. The projecting tube had gone underneath the safety strap on my right ski.

Incredibly, it held long enough for me to be suspended upside down as the chair moved inexorably toward the large rotating wheel above the hut.

The attendant watched, mouth agape, and did nothing as I was lifted toward him.

Finally, the strap broke, and my ski and I fell to the path. I landed on my shoulders on the side of the path. My ski carried on down the hill. Ernie made a heroic dive and caught it.

Even after I had put on the ski (minus one strap), the attendant was still paralyzed in the same position, with his Gauloise stuck to his lower lip, close to burning him. He never stopped the lift.

I decided not to go skiing with Ernie again.

Entertainment in Dubna

The first or maybe second IAEA Large Tokamak Meeting was held in the summer of 1975 at Dubna, a scientific town 70 miles north of Moscow. Because Alan Gibson was working at Princeton that summer, I went as the JET physics representative, along with Paul Rebut and the heads of the other JET departments. Similar groups turned up from the Japanese JT-60, Russian T-20, and U.S. TFTR teams, along with senior people from the various countries, including Tihiro Ohkawa, inventive head of the fusion program at General Atomics in San Diego (Figure 8.2).

After we had all assembled in Moscow, we were taken by bus to Dubna. Winter wreaks havoc on the Russian roads, and the one to Dubna had frequent potholes. All vehicles drove in the center of the road until forced to avoid a head-on collision. The bus—visualize an ancient American school bus—had a questionable suspension. Sitting on the hard bench seats became so uncomfortable during the 2 hour drive that many of us stood and held onto the back of the seat in front, or the luggage rack above, to avoid severe spinal injury. Unwisely, Paul Rebut remained seated behind the rear wheels. As we rounded a bend and just missed hitting a large truck (whose driver did not blink), we saw a sign that unmistakably read, "Dangerous Bump."

"They've got to be kidding," said one of the Americans. Most of us stood and grasped our support structures.

As we hit a monster pothole, the sound "Oof," emanated from the back of the bus.

I turned in time to see Paul Rebut, with a startled look on his face, rise above the seat back in a Buddha-like pose. For the rest of the ride, he stood.

The meeting was exciting and fascinating. Tihiro Ohkawa furnished light relief at lunch, when he would order wine and amuse himself by chatting up the hotel staff. One of his running lines to our young waiter was that, being Japanese, Ohkawa was expert in judo.

The waiter listened impassively. Then, after 2 days of being offered lessons in the art, he said, "Okay. You show me."

Figure 8.2 Attendees at the Dubna meeting outside our hotel.

Tihiro, who customarily wore a neat gray suit, white shirt, tie, and sneakers, stood up and unwisely got hold of the Russian waiter's arms, and said, "Try to throw me."

The athletic young man, who towered over Tihiro, shrugged. He lifted Tihiro off the ground and held him out horizontally.

"Very good," said Tihiro. "Now it's my turn."

The Russian, looking resigned, let go. Tihiro dropped flat on his back onto the parquet floor.

Somewhat dazed, Tihiro rose to his feet. "I think I've taught you all I know," he said. "How about another bottle of wine to celebrate your graduation?"

One afternoon, Tihiro gave an entertaining talk on a proposed new tokamak to be built at the Oak Ridge National Laboratory in a joint venture with General Atomics.

"We've been considering names for the experiment," he said with a deadpan face. "I suggest calling it the Oak Ridge General Atomics Small Machine." At first, only a few of us got the joke, but then he showed his next slide. The acronym, in large letters, read *ORGASM*. Everybody got that one.

Actually, its real name was Impurities Studies Experiment (ISX). In a strange turn of fate, I would end up responsible for ISX, 3 years later.

At the end of the session, Harold Furth from Princeton told us about the famous Russian paper from Novosibirsk, presented some years earlier at a conference, with

Figure 8.3 What is Tihiro Ohkawa (shirt on, for the Russia/U.S. team) trying to do to me (shirt off, for Europe/Japan)?

the coauthor V.I. Foreskin. When Harold commented to the laboratory's director that he hadn't come across that scientist, the director laughed. "You don't get it, do you?" he said. "Foreskin. It's a Russian joke."

For entertainment one afternoon, our hosts ferried us out to an island on the nearby lake, the Moscow Sea, for a picnic. They also organized a soccer match: Russia and the United States (with shirts on) versus Europe and Japan (with shirts off). The playing field was an ill-defined patch of grass and shrubs, edged with copses of trees. Players with the ball would disappear from sight behind a copse, only to reappear on a different part of the field. The Russians clearly dominated, notably the imposing figure of Dr. Strelkov, because they knew how to play the game and were fit (Figure 8.3).

9 JET Design: Do It Again, and Again, and . . . ?

Entertaining Times on JET

During my time working on JET, I shared an office with a chatty, bright French physicist named Denis Marty. Denis took advantage of being in England. He rented a thatched cottage for his family and bought a purple Austin Mini. One weekend, he took his family to the Woburn Abbey safari park. The kids were enthralled when one of the baboons, from a pack gathered around a large oak tree, leaped onto the front of their Mini and peered at them through the window. They giggled when the baboon held onto the wing mirrors with its feet, and the windshield wipers with its hands. It stared at them for a moment before lowering its head and biting the chrome washer with its teeth. Then, with one quick movement, it ripped all of the accessories off the car and disappeared behind the tree. Denis had no choice but to drive on, and it was then, as he moved away, that he spotted a huge pile of chrome car fittings at the back of the tree.

In 1975, I enjoyed both good and bad experiences. My monograph, *Plasma Scattering of Electromagnetic Radiation*, was published by Academic Press. It sold out over the next few years. In 1978, it was translated into Russian and published in the Soviet Union by Atomidzat. My share of their payment to Academic Press paid for a skiing holiday in France for our family.

Mathew Snykers, a Belgian colleague on JET, was facing a problem. He had been sent a sample of an improved form of graphite and wanted to know what kind of texture it had inside. In particular, he wanted to cut it open while messing up the fresh surface as little as possible. He knew that there was a powerful laser at the high-temperature gas-cooled reactor site at the Winfrith laboratory, and he wrote to its director, asking if they would use the laser to cut his sample in half. Unfortunately, he did not explain clearly what he was after.

Three weeks later he received a masterfully penned letter. This is my attempt to recreate it.

Dear Doctor Snykers

I have read your request and have no idea why you need to use our laser to achieve your objective. Nevertheless, your request got me thinking about techniques for meeting your goal. After considerable thought and discussions with my laboratory's metallurgists and engineers, I have come up with the following invention.

Fun in Fusion Research. DOI: http://dx.doi.org/10.1016/B978-0-12-407793-5.00009-1

I propose that one would take a thin strip of metal, steel for example, and make serrations along one of the edges. If this piece of metal were drawn back and forth across your graphite sample, I believe that eventually it would make a thin groove and ultimately it would cut the sample in two.

It seems to me that such a new invention deserves a name. I have chosen to combine the Norse word *hacke*, to rend asunder, and the Saxon word *sawe*, to sever or cut. Therefore I have named the device a *hacke sawe*.

I suggest that you make one and use it.

Yours sincerely . . . .

We completed the design of JET on schedule in 1975 and submitted it to the European Commission. Unfortunately, our concern that the Council of Ministers would not be ready to commit to construction was confirmed when, not long after we submitted the design, our director, Paul Rebut, called us to a meeting.

"I'm sorry to tell you that the rumors are correct," he said. "The ministers can't agree on how to proceed, and we've been asked to refine the design."

"How long will we have to be doing this?" somebody asked.

"I expect another year or so," Paul answered. "They want us to revisit everything."

(In fact, the decision was reasonable because technical issues remained that needed to be resolved.)

A word, muttered in at least six languages, reverberated around the room—Shit!

"One good thing is that they have agreed to let the contributing countries go ahead with site proposals for the project. Each proposal will include discussions of power availability, skilled labor availability, housing, language, schools, and ambience. Euratom will issue a report next year."

We went back to work and refined the design. Because we now had more time to do the underpinning research and development, the design was considerably improved. Yet, well into 1976, we had still not received approval, but we had received a document describing candidate sites—Grenoble and Cadarache in France; Julich and Munich in Germany; ISPRA, an Euratom establishment in Italy; and Culham. During a coffee break, we discussed the site evaluations.

"How about this British proposal?" an English colleague asked. "Listen to the answer on language: 'There will be no problem with language.' Of course not. You go into your local butcher and ask for a cut of veal, it won't matter what language you use. You'll get the same response. 'Wot?' Everyone will have to understand and speak English, like we do on the project."

"I like the French proposal for Grenoble," I added. "Listen to this, 'Grenoble is close to three international airports.' You are all wondering which ones, aren't you? Okay, Lyon is just down the road, but the airport's small. You can get to Geneva, going over lower passes in the Alps. But listing Milan is a bit of a stretch. The Alps, in all of their majesty, get in the way."

(Note that this was before a tunnel was built.)

The German site near Munich was picked on next. An Italian colleague said, "I can't believe this, 'Housing is widely available.' I used to work there. The only place more expensive is Switzerland. It should say, 'Only expensive rentals are possible.' "

"What about the proposal for the Italian site—ISPRA?" a German colleague asked.

To our surprise, our Italian colleague commented, "Listen to the answer on electrical power: 'ISPRA is connected to a strong international grid with a pumped storage system.' True, but what it doesn't say is that there was a study of what our project would do to this grid. It's rubbish."

We all looked at him expectantly.

"At a minimum, TVs would flicker badly in at least four countries every time we operate: northern Italy, Austria, Yugoslavia, and a part of Switzerland ... and even possibly reaching into France and Spain."

I finished my coffee, hoping that our project and the rest of the efforts to find a site would be more competent than this misleading piece of work. Of course, for a number of the countries, the site offers were intended to be bargaining chips for future European facilities—such as the meteorological center. The exercise was mainly political.

A year later, Paul assembled us again to review the project status. "Bad news," he said. "The commission wants another delay. I've expressed in the strongest terms what I'm sure we all feel. In particular, I warned that it might be difficult to hold our team together unless there is a clear statement that the project will proceed. This seems to have made an impression. The Science Commissioner, Dr. Schuster, Donato Palumbo (head of the Euratom program), and the directors of many of your laboratories will come here next week to explain the delay to us."

This was a truly amazing event. We were under the impression that commissioners were probably only second to the pope in the eyes of God—possibly even first.

The following Tuesday, we assembled, with great anticipation, in our meeting room. The dignitaries and Paul Rebut sat in a semicircle on a low stage. The rest of us, about 45 people, sat on rows of chairs in front of them.

Paul stood. "We are honored that the Commissioner has found time in his busy schedule to come and talk to us."

The Commissioner, Herr Doctor Professor Schuster, a tall, immaculately dressed, tough-looking German, looked at his watch and gave a professional smile. "Chaps," he said, in an avuncular tone, "I know how you feel. It is no simple matter to obtain the agreement for a major project like yours. I'm sure you have all seen the outstanding site document." He looked around. "It will be difficult for us to choose among such excellent sites."

He carried on for 20 minutes, basically saying that we could expect further delays.

We looked at each other with raised eyebrows: another snow job.

He finished with, "So, chaps, we're getting close, but we still do not have a timescale. Hang in there and take this opportunity to further refine the design."

When he sat, Paul stood again. "The Commissioner has graciously agreed to answer a few questions."

A long silence ensued, while we looked at each other. Nobody appeared game to speak, but I could see that Denis was fidgeting. When it looked like no one was going to say anything, he raised his hand. "I have a comment."

Paul Rebut said, "This is Dr. Marty, one of our French colleagues. Please go ahead."

Denis stood. His nervousness dissipated as he spoke. "My English is not so good, but I believe that there is a word in the English language that describes what we have been doing. I mean, doing something repeatedly, but not for the purpose for which it was intended."

The dignitaries looked both puzzled and curious about what the word might be.

Denis smiled nervously. "That word is *masturbation*!" he said.

A silence followed while Denis's word echoed around the room. The Commissioner looked stunned. Paul Rebut, his eyes bulging, gaped at Denis. The head of the European program, Donato Palumbo, nearly fell off his chair trying not to laugh. Many of the directors were taking the opportunity to use handkerchiefs to hide their expressions.

"I am tired of masturbating this design." Denis paused. "I want to do the real thing . . . construction."

We clapped and cheered. Someone sitting behind me whispered, "Denis should have said *erection*." The dignitaries stopped trying to hide their amusement.

Of course, the Commissioner had the last word at the meeting. "An unusual way to make the point, Dr. Marty. But it is well taken. We will do our best. Now, chaps, I must get back to Brussels and try and expedite your project."

As a postscript, I should add that it took an international incident, when British commandos freed hostages from a German plane in Somalia, to pave the way for a site decision. You see, the following week, the British prime minister, James Callahan, was visiting the German president, Helmut Schmidt, in Bonn. It is said that the conversation went like this.

"Jim, how can I thank you for what your commandos did?"

Callahan had been primed with an answer. "Let us have the JET project, Helmut (Figure 9.1)."

A strange beginning, but, after various other machinations, this tremendous project was built at Culham. The construction was completed on schedule, with the first plasma achieved in early 1983. The cost overrun of only 8% was a fine achievement, given that this was a period of massive inflation (Figure 9.2).

Shocking Experiences

At Culham, HV activities still played an important role in the 1970s. We had the three neutral beam development groups and also the lightning strike facility. HV in research can be a shocking experience, as discovered by a pasty-faced, young

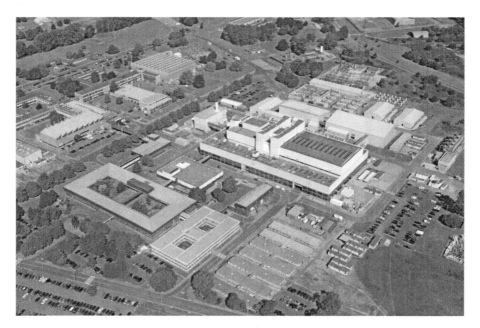

Figure 9.1 Aerial view of the JET site. The left side of the large building (top-center) houses the experiment.
Source: Courtesy of the JET organization.

Figure 9.2 An interior view of the vacuum vessel, showing the glow from a plasma discharge on the right.
Source: Courtesy of JET.

technician who I'll call Kevin. Kevin worked for a senior experimental officer, "Dick" Barton, who had a deadpan sense of humor. (I'm not sure if his name really was Dick, for, during the 1940s, BBC radio aired an adventure program called "Dick Barton Special Agent," a favorite of mine.) One day, I went to Dick's lab to get his advice on some hardware. Kevin was sitting at a bench, eating greasy fish and chips from a crumpled piece of brown paper.

After I asked my question, Dick reached over Kevin, said, "Excuse me," pushed the chips aside, pulled a pen from his pocket, and made a sketch on the paper.

"Does that answer your question?" Dick asked, handing me the torn corner of Kevin's lunch wrapper.

"Yes, thanks," I said.

Dick smiled, "Thank you, Kevin."

The whole time, Kevin had acted as if we were not there. Head down, he had continued munching away at his lunch throughout the exchange.

Later in the 1970s, I visited a foreign laboratory to learn about their neutral beam development program. One of their researchers took me to see the test stand for the 20 kV source that they were developing. As we entered their lab, a graduate student fired a beam and then turned off what I assumed was the main control switch. While his supervisor explained the setup, I watched the student go over to the HV enclosure. To my surprise, instead of opening what appeared to be the safety-interlocked door—which would have disabled the system—he pulled a screwdriver from his back pocket and took off the wire-mesh panel behind the source. Having such screws accessible on the outside was a no-no at Culham.

To my amazement, his boss said nothing. Before I could comment that this looked unwise, the student reached in to make an adjustment to the source, which was mounted on insulators above a metal table. At that moment, the source fired again. The discharge ran through the student's arms to his stomach, where it arced to the metal table. He was hurled back toward me, screaming, "I love my family. I don't want to die."

His head hit the concrete floor. He lay on the floor, ashen-gray-faced and unconscious. His supervisor called for an ambulance, and somebody hurried me away to another facility. I am happy to say that, other than the burns on his hands and stomach, the student was uninjured. I heard later that safety precautions were substantially improved after that near-disaster.

Visiting the Former Soviet Union

Harold Furth of the Princeton Plasma Physics Lab (PPPL) told me the following story about flying with Mel Gottlieb, then-director of PPPL, in 1968 from Moscow to Novosibirsk for a conference. The Tupolev airliner's luggage rack consisted of heavy gauge netting strung along struts above the seats. Mel was in the forward cabin and Harold in the back half of the plane. In the waiting area, Harold had noticed a family of three eastern Russians—father, mother, and daughter—Mongols, wearing their native costume with beautifully embroidered shirts. All three were stern-faced. As he went to his aisle seat, Harold saw them in the row behind and nodded. They remained sitting, with nary a smile. Their belongings, including a wide-brimmed straw hat, were on the rack above him, on which he stowed his jacket.

About an hour into the flight, Harold heard a plop. He looked around, but saw nothing. After a second plop, he looked up and saw a large jar with red jelly oozing from it through the netting. He stood and discovered blobs of jelly on his shirt. He glanced at the family. They looked back impassively. Harold rang for a flight attendant. She decided that the situation needed a more senior hand and called the co-pilot. He came down the aisle and took action. In one sweeping movement, he pulled the straw hat of the rack, not realizing that the brim was full of jelly. The jelly sprayed over the family's ornate shirts. For a moment, they looked shocked, but then they burst into laughter.

Harold, laughing and clutching his jelly-covered shirt front, started up the aisle to tell Mel what had happened. A second flight attendant came through the dividing curtain and, seeing this man who had obviously been knifed, rushed away to inform the captain. Harold returned to his seat and the three Mongols, who grinned at him sheepishly.

Igor Sviatoslavsky works at the University of Wisconsin—Madison. He has the interesting background that his native language was Russian and he spent part of his childhood in Palestine. Like many of us in the fusion energy area, Igor belongs to Fusion Power Associates, an organization that provides educational material on progress in fusion energy development. Every year, they sent members a calendar, and, on one occasion, they included a nifty, telescopic metal pointer. He always carried it with him to use when he gave a talk.

On one occasion, Igor attended a fusion meeting in Moscow. The conference hotel was the huge Rossiya, near Red Square. This hotel had four separate wings forming a square, and each side was assigned to people from a different group—the West, Soviet Union, other Communist countries, and the rest of the world. Although it was possible for a person in one wing to visit other wings, the management discouraged it.

On the first day, Igor was waiting in line for dinner when a tall, heavy-set Russian came into the room and stood near him. Igor found out later that the man was with a group of Russian veterans of World War II.

The Russian had been drinking, and, in his native tongue, he shouted in a slurred voice, "Look at you foreigners, jumping in front of us native Russians, taking all the best food. I didn't fight at Stalingrad to be treated like this."

In fact, nobody in the queue had pushed in. The foreigners all had joined in an orderly fashion. Nevertheless, the Russian elbowed Igor out of the way and stood in front of him, occasionally turning to mutter more imprecations. "If I catch any of you alone, I'll show you whose country this is. We won the war."

Igor didn't let on that he understood everything, and he tried to avoid making eye contact.

During the next 2 days, he saw the Russian occasionally. Every time, the veteran would point his finger at Igor and other Westerners he recognized and complain about them with comments such as "It's not fair, you people coming here and taking our food. You foreigners want to destroy our beloved Russia."

His continual heckling was irritating but not a problem, until the day when Igor tried to take an elevator down to the lobby and none of them was working. He had no choice but to use the stairs. As was the case in most hotels, while you could enter the stairwell anywhere, you could exit only on the ground floor. As my mate approached the third floor, he was horrified to see the Russian standing on the landing. Bleary-eyed, the Russian looked at Igor, and a smile slowly worked its way up his face.

"I knew I'd catch up with you, you filthy American bastard. I'll show you whose country this is." The Russian started up the stairs towards him.

Thinking fast, Igor pulled out his Fusion Power Associates pointer, extended it, and pointed it at the man. In fluent Russian, he said, "I'm with the CIA. This is a new weapon. If you don't get out of my way, I'll have to kill you."

The Russian, showing amazement at hearing his language spoken perfectly, backed away, raising his hands. Igor rushed past him and escaped to the ground floor.

After 3 days at the 1975 Dubna meeting and buying a lot of booze, most of us were running out of cash. Because there was no local facility to exchange traveler's checks or foreign currency, our hosts arranged for representatives from a bank in Moscow to come to our hotel. The next day, two young women arrived with money. They were accompanied by a stern-faced, Kalashnikov-toting guard.

The first person in line, Roberto Andreani, handed over brand-new 100 lire Italian bills. The main cashier, looking nervous, reached under the table and pulled out an inches-thick, hardcover book. She studied the index before turning to a page.

"I guess that gives the exchange rates," someone at the back of the queue said.

A tall man at the front of the line glanced at the tome and replied, "Too thick for just that," he replied, pointing at the woman, who had taken the first bill and was comparing it line by line with an image on the open page.

The tall man craned his neck to see better. "She's got pictures of a lire note," he said. "Front and back."

The woman appeared to be satisfied until she raised the bill to the light. She pushed all the money back across the table, saying something to her colleague. Her expression was a mixture of worry and determination.

"Is forgery," her colleague stated.

"What's the problem?" Roberto asked.

"The watermark is incorrect."

"But I just got them from my bank," Roberto expostulated. "Wait a second . . . what date is your book?"

"The second woman turned the book and showed the front page.

"Oh, 1972," Roberto exclaimed. "Your book's out of date."

The second woman shrugged and said, "Next."

"Good luck," said Roberto with a laugh as a German scientist proffered deutschmarks.

He was successful, but the next person, offering U.S. dollars, failed because the notes were signed by the new Secretary of the Treasury. And so it went on, with various currencies and traveler's checks being declared forgeries because they had the latest signatures, artwork, or watermarks.

When it was my turn, I handed over my five-pound notes with trepidation: I could see that the Russian book's governor of the Bank of England was not the one whose signature was on my bills. Fortunately, the cashier was getting tired of tracing out each curlicue. She was sufficiently impressed with the cunningly interwoven, security metal band in the notes that she forked over my rubles. After 2 hours, less than half of the attempted exchanges of deutschmarks, yen, francs, and traveler's checks had been successful. This bureaucratic experience clarified why, despite the Russians being charming hosts, the Soviet Union was nonfunctional.

How things changed over the next decades. At a meeting some years later, a group of us were trying to make purchases with rubles. The cashier became more and more irritated with us, making comments that we didn't understand. Finally, she reached under the counter and extracted one of those credit card devices that took an imprint of the card.

"She wants credit cards, not rubles," someone said. "Foreign currency!"

The woman held up the device and drew the imprinter back and forward furiously, as if she were playing a musical instrument. We handed over our cards. She smiled triumphantly.

The Soviets were also paranoid about people bringing in goods to sell illegally. This presented a problem for one colleague because he was used to bringing presents for the people he worked with. Each time, he would struggle to find something beyond cigarettes for the men and stockings for the women. On one trip, he came armed with the most useful of presents: plastic raincoats, the kind that come in a small zip-up bag. Gerald had something like 20 of them in his suitcase, a mixture of men's and women's sizes. Unfortunately, when he went through customs at the airport in Moscow, the officer decided to inspect his luggage.

"Ah, ha," the officer said, or whatever the Russian equivalent is, followed in broken English by, "Is illegal bring goods to sell. I take and you pay fine." He called his superior over and showed him the contraband.

My friend had to think fast. He couldn't say that these were presents: It would sound like bribes. "Wait a minute," he said. "These are for me to wear when it rains. They're a new American invention: disposable raincoats. You only wear them once, and then you throw them away."

The two Russians looked startled. But they bought his story: American society was utterly wasteful. Shaking their heads, they indicated that he could pack and move on.

10 1977: Back in the U.S.A.

In late 1976, I visited the Oak Ridge National Laboratory (ORNL) in Tennessee to learn about their neutral beam development work. The government reservation in Oak Ridge, appropriated in the 1940s as part of the Manhattan Project to build the bomb, covers some 50 square miles. The facility sites are labeled by a letter and a number using a grid pattern that remains obscure to me. The main part of ORNL is at X-10. For historical reasons, the Fusion Energy Division was situated in the east end of the Y-12 area. Y-12 is part of the DOE nuclear weapons complex that makes bomb components. The Fusion Division was housed in an ancient building that originally had contained calutrons for separating out uranium 235.

After my tour, Bill Morgan, the deputy division director, asked if I would be interested in moving to ORNL. I said that I might be. John Clarke, the division director, indicated that they would like me to head up their new tokamak project, called ORMAK Upgrade. I was frustrated with the slow progress in selecting a site for JET and looked forward to managing a project, and my wife agreed that we should return to the United States.

We arrived in Oak Ridge at the end of August 1977. The reality of our situation turned out to be somewhat different from what I had expected. Well before September, the ORMAK Upgrade project had been canceled. I felt very uneasy and began to regret the move from a secure job at Culham.

In place of ORMAK Upgrade, ORNL had obtained an agreement to rebuild the Impurities Studies Experiment (now called ISX-A) as ISX-B, the purpose being to study beta limits to tokamak performance. Beta, the ratio of plasma pressure to magnetic pressure—a measure of fusion power over cost—is a key indicator of viability for fusion power generation. Numerous studies show that for economic reasons, beta should be 5% or more, or else the cost of the magnets would be prohibitive.

By November 1977, John Clarke had left ORNL to become deputy director of DOE's Office of Fusion Energy. Bill Morgan became the division director, and Lee Berry became the fusion program director. I was appointed to be head of the tokamak section, which included ISX-A (coming online in December 1977) and the ISX-B components in construction.

ISX-A had a number of subprograms with commitments to not only the prime collaborators from General Atomics (GA) but also the many from other institutions. With the commitment to install ISX-B, we would have only 12 weeks of operating time to accomplish our goals. I was very conscious of the fact that this was an awkward situation, with many people feeling that they had been let down. I worked

Fun in Fusion Research. DOI: http://dx.doi.org/10.1016/B978-0-12-407793-5.00010-8

with the senior staff and our collaborators to make the best of the situation. As a first step, I set up a joint management of the program under Tom Jernigan for ORNL and Keith Burrell for GA. I also talked to the team and suggested a flat-out campaign. We decided to work two shifts, 5 days a week plus 1 day for maintenance.

By the time ISX-A shut down, the team had accomplished essentially all of its goals, including the study of impurity flow reversal in the GA part, a suggestion of Tihiro Ohkawa. Highlights of the other subprograms were the first injection of solid hydrogen pellets for refueling, with laser holograms of the ablating pellets produced in collaboration with MIT; and testing of a variety of material limiters for the plasma.

Security—A Serious Matter

Yes, security was a serious matter, even in the unclassified ORNL part of Y-12. Working there was a strange experience after Culham, where the security was modest in scope (except for the patrol force shooting the occasional errant briefcase). I did not expect a lot of humor from my new situation. But even in Y-12, there were opportunities to giggle.

Don Batchelor, currently head of the theory group, played the trombone in a local big band. One day, a friend brought him a stack of 1940s big band records. When he passed through the guard post that evening, a guard confiscated the records.

"Why are you taking them?" Don asked.

"It's recording material, sir." The guard replied. "See the rule posted on the wall? No removal of recording equipment without a permit."

"But it's not recording material. They're *recorded* material."

"Recording material stays here."

"They're from the 1940s, for God's sake!"

"If you continue swearing, I'll have to call the captain."

It took a number of days calling security to get the albums released.

Then, the issue of where to wear your badge came up. Detailed instructions were sent out defining the appropriate region as being above the waist and below the neck. One of our guys got so irritated with the bureaucratic writing that he pinned the badge to the backside of his pants. When the guard asked to see his badge, he turned, bent down, and lifted his jacket. The guard reported him, and our director had to put a comment in his annual review.

My favorite story involves the time when a colleague was dropped off at the guard house by his son. They are an affectionate family. His son gave him a big hug. A guard with no sense of humor reported them for engaging in homosexual activity. What a joke!

Writing about security guards reminds me of a story I heard about the time a scientist in Europe, a burly, middle-aged Swede with thinning blond hair, visited the Fontenay aux Rose research center in Paris. In France, if you are a foreigner, you have to hand in your passport when you arrive at any government laboratory and pick it up when you leave.

The visitor went up to the desk and presented his passport. "I'm here to see Dr. Lavalle."

The guard inspected the passport carefully. "It doesn't look much like you."

"It's an old photo. I was younger and thinner then. I had more hair."

"It's not a good resemblance."

"Let's face it. I've changed. We all do."

The guard looked unimpressed. "Quite a lot, I'd say."

My colleague shrugged. "That's the way it goes."

"You're Swedish?"

"Yes. Like my passport."

"I see." The guard still looked unpersuaded, but he picked up the phone and called Lavalle's office.

When my friend left, he recovered his passport. Out of curiosity, still puzzled by the guard's reaction, he opened it and looked at the photograph. An attractive woman with long blond hair looked back at him . . . his wife!

The Elmo Bumpy Torus

As I mentioned before, the main problem with a simple magnetic mirror is that losses from the ends make it impractical to produce a hot plasma efficiently in any reasonable length of the system. Even a 100-yard length would be inadequate. Various solutions have been proposed over the years, including the tandem mirror, which uses special end plugs that act to contain ions and electrons, respectively. Another option considered was to arrange the mirrors end to end in a circle. The issue with this approach is that the plasma in the individual simple mirrors is unstable, and problems remain with the curved regions in the toroidal direction (Figure 10.1).

Prior to arriving in Oak Ridge, I knew little about the unusually named Elmo Bumpy Torus (EBT). This configuration consisted of a toroidally linked set of evacuated cavities, each containing a simple magnetic mirror field and having a clever approach to providing stable confinement. It was the invention of Ray Dandl, a creative electrical engineer in the ORNL Fusion Energy Division. The EBT was powered in steady state by microwaves. Each cavity became a microwave oven in which the microwaves produced plasma and an energetic ring of electrons. The goal was to make the current in the rings large enough that it would reverse the magnetic field in the cavity, leading to excellent confinement of the charged particles.

I was told that Ray gave various answers when asked the question, "What does Elmo stand for?"

"Maybe Saint Elmo's fire."

"Could be Elmo in the *Blondie* cartoon strip."

Figure 10.1 The EBT.
Source: Courtesy of the Oak Ridge National Laboratory.

"Possibly a smart uncle of mine."

We will never know.

A very important legacy of Ray's research was that the needs of the upgraded EBT-S drove the DOE to invest in the development of a higher frequency, steady-state source of microwaves: the gyrotron (the Russian success mentioned before). ORNL obtained a contract with the Varian Corporation to develop 28 GHz gyrotrons. Odell Eason, a gifted microwave engineer who worked with Ray, managed the contract. The first gyrotrons delivered were used not only to improve the performance of EBT-S but also to assist the start-up of the current in ISX. The contract was then extended to include the development of higher frequency gyrotrons (60 to 80 GHz) for a proposed bigger bumpy torus, EBT-P, and also for more general application in the fusion program. Since that time, gyrotrons have been developed at up to 170 GHz for use in the ITER; more about this later.

Support for the EBT-P came when Ray and his team won an Office of Fusion Energy (OFE) competition for a new, innovative fusion project. The win came with a condition that the private sector would be able to bid for the project, both with regard to design and with regard to the management and the site. Dissatisfied with what he felt was inadequate support from ORNL for his work, Ray resigned in 1979, set up his own company in California, and joined the team of one of the four industrial consortia bidding for the $100 million—scale project, TRW. The other bidders were Grumman, McDonnell Douglas, and Westinghouse. I was given responsibility for the EBT experimental program.

Because there was a measure of uncertainty about the EBT's performance, one of my first acts was to review all of the experimental data. As far as I could tell, the data was consistent with the model of performance that had been developed; however, the electron temperature data, based upon electron cyclotron emission measurements, was difficult to interpret. Specifically, the data showed three temperature regions, at a few electron volts, around 100 eV, and in the kiloelectron volt range, respectively. The experimentalists' interpretation was that the lowest temperature reflected particles interacting with the wall. The highest temperature was related to the energetic electron rings; and the middle temperature was that of the bulk plasma. Our DOE sponsor, Bill Ellis, was very supportive of resolving EBT issues and increased our experimental budget considerably (two to three times) so that we could resolve this and other uncertainties in preparation for EBT-P. Most important was that we obtained funds to install a Thomson scattering system to measure the electron temperature.

Upgrades to the EBT occurred over the next 2 years, and we undertook our own design effort of an EBT-P, to be sited in our main fusion building in Y-12. Nermin Uckan, a theoretical physicist from Turkey and an EBT expert, was brought into the activity. She played a key role in establishing the ORNL design.

Meanwhile, an experienced project manager, Al Boch, was assigned to handle the competition and run the ORNL part of the EBT-P. Al had been project manager for the nuclear merchant ship, the *Savannah*, and also for the High-Flux Isotope Reactor at ORNL. At the end of all of the deliberations and analysis of the proposals, the competition was won with an excellent design from McDonnell Douglas. I suspect that a key factor in their win was their purchase of land outside the Oak Ridge Reservation, a mile to the east of Y-12. The scale of the activity had now become so large that ORNL management decided to reestablish a separate section for EBT-S, and Lee Berry was put in charge. I was given a new role as associate division director, responsible for tokamaks, theory, and work carried out in other ORNL divisions: atomic physics, materials, neutronics, and robotics.

Al Boch was a feisty program manager who, while he was likeable, nevertheless could stir up strong emotions. He was particularly tough on our senior colleagues at McDonnell Douglas. The Christmas after an initial budget was approved, work started on the EBT-P project, and Al received a card from a McDonnell Douglas manager. It read, "Season's greetings and f*** you."

Data from EBT-S using the new diagnostics came in over the next months, notable preliminary data from the new Thomson scattering system. Lee Berry and Dave Swain, who had transferred from the ISX-B program, soon realized that this electron temperature data did not support the previous interpretation of the electron cyclotron emission diagnostic. The bulk temperature was the lowest one, a few electron volts. The higher temperature parts were only for a small number of electrons. In addition, reanalysis of the data on the energetic electron rings showed that they were not sufficiently powerful to provide a magnetic well. Consequently, the energy put in by the microwaves was rapidly leaking out of the plasma.

The timing was excruciatingly awkward. OFE was about to submit the full budget for EBT-P to Congress for approval. Murray Rosenthal, ORNL Associate

Laboratory Director, called a meeting of Lee Berry, Bill Morgan, and me. Lee stated that the evidence from EBT-S was clear; as proposed, the EBT-P wouldn't work. Murray immediately called the OFE and McDonnell Douglas. To say that they were taken aback does not quite describe their emotional reaction. Within a day, OFE withheld the budget submission and initiated a review of the EBT program. The conclusion of the review was in agreement with the ORNL assessment that the EBT did not work as anticipated. While ORNL scientists developed a new version that would overcome the problems of the existing device—the EBT-Square—interest in this line of research had dissipated and the area was eliminated in 1985 (see the Fusion Dinosaur Chart a bit later).

I have one other memory of Ray. We were sitting in a Fusion Division management meeting when Bill Morgan mentioned that Y-12 had informed him that they had received toilet paper that was slightly too wide, and it wouldn't unroll in the toilet paper dispensers.

Ray, who was sitting next to me, leaned across, and whispered, "It wasn't a mistake, you know. Y-12 did it on purpose. They've always been out to get us." To this day, I don't know whether he was joking.

The Early 1980s

By the early 1980s, the tokamak program was going well, following the success of the Princeton Large Tokamak (PLT) in achieving a central ion temperature of 7.5 keV (80 million °C), the achievement of 3% beta in a number of tokamaks, and the recovery from the L-mode owing to the work in Asdex. The operation of the large tokamaks, JET, JT-60, and TFTR, was anticipated with excitement.

Listening to the Father of the Hydrogen Bomb

The first part of the 1980s was a time of stress for the Clinch River Breeder Reactor, an advanced fission reactor to be built on the Oak Ridge Reservation. (In the long term, as the small part of uranium that is fissile, U-235, runs out, breeders will be needed to convert natural uranium, U-238, to fissile plutonium.)

The phone message from Bill Morgan was brief. "John, I have an interesting opportunity for you. Edward Teller will be in Knoxville for a conference and would like to hear about our fusion program. I suggest that you and Mike Saltmarsh meet with him. Okay?"

"Sure," I replied quickly. "Will he come to the lab?"

"No. The meeting's in his suite at the Hyatt. He's set aside an hour."

Wow, I thought, Edward Teller, "the father of the hydrogen bomb," advisor to presidents, and a regular visitor to Congress. Years earlier, I had heard him lecture at the University of Texas. I remembered this impressive speaker with a thick Hungarian accent, and sharp eyes peering out from under bushy eyebrows. He had looked hard at some member of the audience as he made each point, the victims nodding in agreement like bobble-heads in the back of a car.

"Which day?"

"Wednesday at ten." Bill laughed. "It should be interesting. He comes with an entourage, and management from the Clinch River Breeder Reactor will also be there."

"Clinch River ... what for?"

"Sorry. I should have mentioned it," Bill said. "Teller's promoting his fusion—fission hybrid as an alternative to fission breeders, such as the Clinch."

"You're joking," I replied, realizing what a politically sensitive issue Teller's claims were. "Fusion's not ready for prime time yet ... and certainly not the mirror. At least the Clinch breeder can guarantee making nuclear fuel."

"Just be careful what you say," Bill said. "You'll tell Mike?"

"Right away."

I rushed off to Mike's office, eager to get his reaction to our "opportunity."

"Sounds like fun. What do you want me to do?" he asked.

"How about I do the experiments and theory, and you do the technology and the breeder? You've more experience in fission."

Mike made a face. "Thanks a lot, I guess. The Clinch River people can't be happy."

Wednesday found us waiting nervously outside Teller's suite while he finished talking to another group. As the group filed out, one of them patted me on the shoulder and smiled commiseratively.

The physical setup in the reception room was simple; however, the crowd arrangements were not. Two sofas were separated by a coffee table. Edward Teller sat in the middle of the one facing the door. After a round of introductions, Mike and I sat opposite him. Behind us, the management of the Clinch River project lurked. Behind Teller were the meeting organizers, and two Air Force colonels who ferried him around the country.

Teller offered us coffee, which we accepted. "Now, tell me about your Oak Ridge fusion program." His eyebrows expressed great interest.

I started with a summary of the principal elements of our program and had not quite finished when he raised a finger and interrupted me. "Most interesting. Now, what do you know about *my* fusion—fission hybrid?"

"It's based on a tandem mirror, I believe," I replied.

"Good. So you know about it." Teller went on to explain at length why he thought *his* system would be better than a fission breeder, like the one proposed for the Clinch River site. I could feel some of the guys behind us clutching the back of our sofa.

After a while, he paused and said, "I am not being a good host. Let us return to your ORNL fusion program." He rolled the *R* in ORNL like a Victorian gentleman who had practiced, "round the ragged rocks the ragged rascal ran."

I looked at Mike, and he started to describe some of our technology strengths. Mike had just mentioned robotics when Teller raised his finger again.

"I believe that *my* concept lends itself to remote handling," Teller said. "We will maintain the walls through the holes in the end."

Mike and I had agreed that we would avoid argument, but this statement was too much for me. I raised my hand, and Teller indicated that I might speak.

I tried to choose my words carefully. "With all due respect, sir, from my understanding, this tandem mirror will be at least 100 ft long, and the holes at the end only a few inches in diameter. It will be easier to service the walls by taking modules out, sideways."

Silence.

"Interesting point. Does your colleague agree?"

Mike said, "Yes."

Teller turned to the colonel who was taking notes. "You have noted that point, I hope."

The colonel nodded.

"I will convey it to my associates. I am being remiss again. We should return to your program."

I continued where Mike had left off and managed to get through about a quarter of our program. Teller asked insightful questions, but he began to fidget.

When I paused to take a sip of coffee, Teller stopped fidgeting and asked the question that Mike and I dreaded. "Now to the $64,000 question: Which would you choose, the Clinch River Breeder reactor or *my* fusion−fission hybrid?"

Without hesitation, Mike and I said, "The Clinch River breeder."

A long silence followed. The hands gripping the sofa relaxed. A colonel stopped taking notes.

Teller's eyebrows showed disappointment. "A straightforward answer, but of course I disagree with you."

The second colonel tapped Teller on the shoulder and pointed at his watch.

"I fear that we have run out of time. The time we spent discussing my ideas means that I will not be able to hear more about your most interesting fusion program. However, there is one thing you should know." Teller paused to emphasize his next point. "I ... am a very good listener."

Muffled sounds came from behind our sofa. The colonels' faces showed amazement.

Edward Teller smiled and raised his finger. "But only to myself."

Epilogue

The minute we closed the door to Teller's suite, the Clinch River management thanked Mike and me for supporting their project. The senior manager even hugged us. Unfortunately for him, and having nothing to do with Edward Teller's option, Congress terminated the project in 1983. The tandem mirror program didn't survive, either!

11 Conferences in Erice

Erice is perched on top of a 3000 ft mountain at the east end of Sicily, near the coast at Trappani. In mythology, it was the birthplace of Venus. An advertisement in a brochure for the town states that the outer walls were last repaired in 1500 BC. The ruins of a Venetian fortress rest on top of the remains of a Norman castle that in turn sits on top of a Roman structure, which was built on even older foundations. Narrow, cobbled streets wind between the walls of old weathered stone houses. An occasional open lattice metal door allows a glimpse of a tiny garden. A Baroque church dominates one of the squares.

The Ettore Majorana Center for Scientific Culture hosts a number of conferences every year. Prof. Antonino Zichici headed the center and ensured that each meeting would involve the arts as well as the sciences. The meetings are held and most of the accommodation is provided in a building that at one time was a nunnery.

My first opportunity to go to Erice came in 1976, when I was still working on JET and regularly giving talks on what we were planning for heating the JET plasma. One week earlier, I had attended a conference in Varenna on Lake Como. One evening, I had dinner with Harold Furth and two of his colleagues from the Princeton Plasma Physics Laboratory (PPPL). Harold was one of the world leaders in fusion research, and later director of the laboratory. He had contributed important papers on theory and initiated many innovative experiments. I always paid close attention to his views on where the program ought to be heading.

"Let's toss a coin to see who pays for the wine," Harold said.

I was in a happy state by then, and responded, "Between you and me, Harold. Heads or tails, you pay. If it ends up on its edge, I'll pay."

Harold flipped the coin. It landed on the floor, rolled around, and rested upright against the table leg. I never again agreed to a coin flip with Harold.

It turned out that Harold was also going on to Erice, and we agreed to rent a car and see some of the ancient ruins on our way there. We flew to Palermo, arriving early on Saturday morning, and picked up a rental car.

I had started writing a novel and wanted to gather data for some of its settings. Harold kindly agreed to take the route that I found interesting. First, we drove to the south coast to see the incredible Greek temples at Agricento. From there, we continued around the southern coast to Marsala (of fortified wine fame) and Trapani. If we had realized what a rustic road we would be driving on, we would have gone back to Palermo before heading east. We arrived in Trapani, tired and hungry, as the sun was setting. By now, having driven for hours on the tiny, winding road, I had a bad headache. Because our accommodations in Erice were not

Fun in Fusion Research. DOI: http://dx.doi.org/10.1016/B978-0-12-407793-5.00011-X

until the following night, we searched for a hotel in the town. Finally, we found a run-down one in a back street. We checked in and took the elevator up to our rooms on the fourth floor. The corridors were dark and grimy, and the lights, on timers, had to be turned on every few yards. We located Harold's room, and I continued on to mine. The room's tiny window would not open and offered a view of a neighboring building's wall. The floor was covered in white tiles that continued 4 ft up the walls. One aged wardrobe was the only furniture besides the bed and a night table.

I dumped my suitcase on the floor and went to get Harold so that we could go to dinner. He was still standing in his doorway, suitcase at his feet, staring at a room similar to mine except that his window looked out into an airshaft.

"I have a feeling the Mafia uses places like this to deep-six their enemies," he said. "It would be so easy to clean the blood off these tiles."

He went over to the wardrobe and opened it. A stale roll rested on a threadbare extra blanket.

" John, I can't stay here. We've got to find something better."

"Where?" I asked. "We didn't see anything else."

"We're on the coast," Harold said. "There must be tourist hotels. We're checking out. I'm going to ask the desk clerk."

Back in the lobby, Harold went up to the desk. He handed our keys to the clerk. "We have to go," he said. "Where is the nearest tourist hotel . . . *touristico*?"

The clerk looked bemused, but gave a name. Harold took a pen from the desk and indicated that he'd like to see it written down. After the clerk responded, Harold pointed at the telephone and made a dialing sign with his finger. "Can you book two . . . *due* rooms?" he asked.

"You're kidding, Harold," I exclaimed.

"What have we got to lose?"

Amazingly, the clerk did what he was asked, and even showed us on the map where the hotel was situated, north of Trapani on the beach.

As we were driving along the beach road, we passed a sparkling new hotel. "Turn around," said Harold.

"I don't think we've gone far enough," I said.

"If it's the one, fine; if not, we'll see if they've got anything. We can cancel the other one later."

It wasn't our hotel, but they had great rooms available, and a wonderful restaurant. After dinner, we drove a couple of miles up the coast and found the original hotel and canceled our reservation. I would not have had the nerve to do what Harold did, but I was very grateful for his actions. By the time I had my second glass of wine and had started on a fine veal dish, my headache had gone.

One of the attendees at the meeting was Bob Carruthers, a charming man who had been the principal engineer on ZETA. His only fault that I observed was a penchant for waffling at length in a plummy British accent when he gave a talk. This quirk made it hard to understand the points he was making. Nevertheless, when it was his turn to speak, I went to listen. The small lecture hall was in an old building alongside the garden. It had French doors opening onto a patio. When I arrived, the hall was crowded, and I had to stand outside the open doors.

Bob was introduced and he started his talk.

"Well, hm, hm, hm, it is and uh, mister chairman, let me assure you ah, ha, ha, a pleasure to visit your uh, beautiful uh coun ... try and uh, enjoy the uh sights. Eh, eh. I will not use slides uh. It is more hm, hm enjoyable to just uh stand here uh and uh chat. Of course." Chuckle. "We all uh know it, wee-ell what can I say uh, is uh really uh, hm, hm, hm a bit of a aha, ha, joke to get ... uh ... paid for this um job in ... er ... fusion energy."

After 10 minutes of this incomprehensible talk, a Yugoslav standing next to me muttered, "I don't understand a word."

"I do," an Italian on his other side said.

"You do?" The Yugoslav's face showed his genuine amazement.

"Perfectly. Is exactly the same in Italian. Hm, hm, ah ha ha. Prego, eh, eh, pasta, uh." Chuckle. "Ha, ah, ha, um spaghetti-ah."

It was a good thing that our little group was outside the hall. The speaker never heard our chuckles.

My second visit to Erice was in 1985, for a conference titled "Tokamak Start-up: Problems and Scenarios Related to the Transient Phases of a Thermonuclear Fusion Reactor." I gave the first paper, summarizing progress on this topic in the world program. My colleague, Nermin Uckan, followed with a talk on "A Simple Procedure for Establishing Ignition Conditions." (This is the point at which the alpha particles from the fusion reactions begin to provide enough power to maintain the plasma temperature.)

The conference was admirably organized by Heinz Knoepfel, a genial Swiss who worked at the Italian Frascati Laboratory. As I mentioned, these meetings also included components from the arts. The first of these was a recital by one of Italy's premier flautists, Maestro Severino Gazzelloni, held one evening in Erice's baroque church. However, his flamboyant performance and golden flute did not impress Susan Pease, Bas Pease's wife, who taught the flute. I remember the way Gazzelloni drew attention to himself when his accompanying pianist, Leonardo Leonardi, was allowed a brief solo. The maestro periodically extended his right arm in a dramatic gesture every time Leonardi played a trill.

But the artistic highlight was a performance by the *folkloristico* group from Marsala, who sang Sicilian sea chanteys before the conference reception and dinner. Picture the scene in the Charlie Brown Christmas cartoon in which all the children stand around the tree and open their mouths wide to sing. Now imagine the same thing with 20 or so adults in a small lecture theater. The songs were salty, invigorating, and very loud.

At the front of the group was a small boy, maybe 7 or 8 years old. His role became clear when, during the third chantey, he inserted something in his mouth and raised his right hand. At the prompting of the lady behind him, he twanged a few chords on a Jew's harp at appropriate points in the song. At the end of this sea chantey, he took a bow to great applause.

It would have been nice if the concert had been briefer. But after 40 minutes, the group was still going strong. We were all ready for the reception and drinks, and the little boy was fidgeting.

Suddenly, while a soprano was belting out a rousing number about her seaman returning, a tiny twang echoed around the hall. I saw the boy drop his hand. The members of the choir couldn't see him, but a few of them glared. Other surreptitious twangs followed until, finally, a large hand closed over the boy's mouth.

The hand remained firmly in place during the next two songs, while the boy squirmed. At last, he managed to wriggle away sufficiently to uncover a part of his mouth. His hand shot up ... twang, twang ... twang, twang, twang, he went, his index finger inserted between two adult fingers. As one, while still singing, the choir turned toward him. Two hands now grasped him firmly for the remaining songs.

"We will now take a break," said the choir leader.

"That was wonderful," said Heinz, quickly. "Let's all give the *folkloristico* group a great hand of applause before going to the reception." We clapped and cheered as we started to leave.

"But, but, we haven't completed our program," protested the leader.

"Unfortunately, the reception cannot wait for us any longer." Heinz looked sympathetic. "You must all join us. It will be a terrific opportunity to mingle and for us to learn more about your work." He grasped the choir leader's arm and drew him, still protesting, out of the hall.

At the reception, I took Heinz aside. "Thank you for a wonderful evening. I enjoyed the singing, but an hour was enough."

"I looked at their complete selection of songs. The second part was going to be longer."

"That poor little boy. What would he have done?"

Heinz grinned. "You know, maybe it would have been worth it."

Heinz did a wonderful job of organizing an entertaining meeting, unflappably dealing with every crisis as he did with the singers. The only time I saw him worried was the day we had an afternoon off. He had organized a cruise for us to the Egadian Islands that sit off the western end of Sicily—Favignana, Levanzo, and Marettimo. After the hydrofoil from Trapani dropped us off at the dock at Favignana, we waited for our boat to appear. No boats were moored at the quay, and Heinz paced nervously, occasionally looking out to sea. After a half hour or so, a strange piratical-looking vessel appeared around the point (Figure 11.1)

Heinz looked concerned.

"Are you afraid this isn't our boat?" I asked.

"No," he replied. "I'm worried that it is."

It was. But the voyage was a lot of fun. The boat took us on a circumnavigation of the neighboring island of Levanzo. On the way, the crew put out a net and caught an odd assortment of little fish. The captain's wife then concocted a wonderful seafood pasta using the catch. The highlight of the trip was a visit to the Grotto di Genovesi, contained Paleolithic art; an isolated cave on the west end of the island at about 100 ft above sea level. An outer room is open to daylight. In

Figure 11.1 The pirate boat off the island of Favignana.

the inner, darker room are incised drawings of cattle, humans wearing birds' head masks. A second group of drawings is painted in red and black and depicts men, violin-shaped women, mammals, and fishes, and one figure appears to be dancing.

A footnote to this most enjoyable conference: I was inspired to write two poems as a commentary about progress and controversies in fusion. To my surprise, Heinz put them in the conference proceedings of the seventh course of the International School of Fusion Reactor Technology, held July 14−20, 1985, in Erice, Sicily.

The Winding Tokamak Road (with apologies to G. K. Chesterton)
Before Ulysses came to Troy
And out to Eryx strode,
The cunning Elimini
Built a winding twisting road.

A winding road, a tortuous road
That rambles round each hollow,
And after them the Romans came,
The Carthaginians and Palumbo.

This route we followed gladly
In hope as we flee
On the way to giant JET
By way of TM-3.

I have no fear of MHD
And for high power I am praying,
But my feet they nearly left the road
When we encountered Goldston scaling.

The cynics they will mock you,
And advise you not to see
That the road to better current drive
Is through the use of higher Zee.

The road it had a mind of its own
It swerved from Moscow
Smartly across the USA,
To wind up in Frascati.

But such a road it is that we will tread,
As the whole truth we will see,
On the road to full ignition
By way of Bruno Coppi.

* The Elimini were the early inhabitants of Monte Erice, the birthplace of Aphrodite. In the 1980s, Donato Palumbo, an astute Sicilian, was the head of the Euratom Fusion Programme, of which the JET tokamak experiment was the centerpiece.

* TM-3 was an earlier Russian tokamak that paved the way to this approach.

* MHD stands for magnetohydrodynamics, the fluidlike behavior of hot fusing material that has the potential to cause the plasma to run amok, but is well controlled in a tokamak.

* Rob Goldston, later director of the PPPL, was one of the developers of simple models of how plasma losses scale with the size of a tokamak.

* Frascati, to the south of Rome, is the Italian government's fusion laboratory.

* Ignition is achieved when the reacting fusion material is hot enough and contained long enough to support burning the fuel.

* Bruno Coppi is a creative Italian professor at MIT. He believes that his approach to the tokamak, using very high field copper coils and called ignitor, is the only one worth pursuing. This is a goal he has pursued with vigor for 30 years.

The Inferneti (with apologies to Lewis Carroll)
T'was Erice and the debate was on,
To ionize and ramp-up from the wave?
All happy were the RF crew,
Though the designers still were grave.

Beware the up-shift and the gap my son,
And from the return current shy.
Beware the booze and shun
Yon dreaded water supply.

He took his launching structure in hand.
Long time the Infernet he sought.
So rested he in Erice,
And sat a while in thought.
And as in uffish thought he sat,

the Infernet with eyes aflame
Came lurking through the stone clad streets.
And burbled as it came.

One two, one two, and through and through
The flashing wave crossed the device,
The current rose, and is it burned,
He went to claim his Nobel prize.

Oh has't though raised the current my son
Come to our arms my beamish boy
For you have minimized poloidal woes,
They chortled in their joy.

T'was Erice and all agreed
To ionize and ramp-up from the wave.
All happy were the RF crew,
For the Infernet was saved.

* RF stands for radio frequency. The RF waves are one approach for heating plasmas.

* NET was the Next European Tokamak, intended to follow JET. "Infernet" is recognition that some scientists view it as an infernal machine. It was superseded by ITER (Latin for "the way"). ITER is under construction in France.

12 The Winding Stellarator Road

During the 1980s, Martin Marietta took over the contract from Union Carbide to run the facilities on the Oak Ridge reservation, and the gaseous diffusion plants at Paducah, Kentucky, and Portsmouth, Ohio. Overall, the change was good. Martin Marietta put far more effort into thanking people for their good work. It also got rid of the ludicrous system in which scientists who patented some invention received an acknowledgment and $1! In the new system, scientists could share in the royalties, and if the Department of Energy (DOE) and the company were not interested in patenting the idea, the scientists could patent it themselves.

Some years earlier, I had discovered how bad the patenting system was. All patent applications had to go through the Oak Ridge Operations Office (ORO) of DOE for approval. Jim Decker, deputy at the DOE Office of Science, asked me to obtain for him data for the previous 10 years on the fate of patent applications from the ORNL Fusion Energy Program.

I found 100 applications, of which only 10 were approved. I called the ORO patent examiner and asked why the success rate was so low.

"The realization of fusion energy is so far off, how can they be relevant?" he replied.

"Because we are developing state-of-the-art technologies in cryogenics, superconductivity, particle beams, microwaves, and so on," I retorted.

"Give me some examples."

"An accelerator for solid frozen pellets of hydrogen. An improved method of joining superconducting cable. The brightest ion beam source—"

"Those are all for fusion—it's decades away."

"Our frozen carbon dioxide pellets are being used to clean paint off F-15 fighters," I responded. "Superconductors are used in magnetic resonance imaging. Ion beam brightness is important for Star Wars." Then I added the clincher. "I hear that scientists in other countries are patenting these ideas."

Silence for a long moment; then he said, "You've made your point. I'll look at them more carefully. Also make sure people submitting patents explain such applications."

* * *

Norman Augustine, who later became head of Lockheed Martin, used to visit all of the Martin Marietta sites, yearly, and explain how the company was doing. After his entertaining review, he would take questions on any subject. The attendees were handed white cards on which to write their questions. The questions that he

Fun in Fusion Research. DOI: http://dx.doi.org/10.1016/B978-0-12-407793-5.00012-1

did not have time to answer were all answered in the site newsletter. I remember one question well:

"I read that when Lockheed and Martin Marietta combined, you received millions of dollars. I also believe that you earn more than a million a year. How do you justify that?"

"As you can imagine, I have been asked that question before," Augustine replied. "In regard to the money from the merger: it was written into my contract. I did ask whether I could turn it down. The answer was no. I gave a lot to charity."

Augustine took a drink of water. "Now, about my salary; when I was a young aeronautical engineer, I could not have conceived that I would make as much as I do. I should point out that our company analyzes the pay of senior executives at peer companies, such as Boeing, Grumman, Northrop, and so on. We pay in the middle of that group." He paused. "One other thing . . . you should note that I make a lot less than Hulk Hogan."

Norman Augustine impressed me enormously with his command of his subject, his straightforwardness, and his wit. I strongly recommend that you read his entertaining memoir about a life in the aviation business, *Augustine's Laws*. His book inspired me to write my memoir.

Stellarators and Koji Uo

My involvement with stellarators was linked closely to the career of a Japanese colleague, the late Professor Koji Uo of Kyoto University.

Princeton University's Lyman Spitzer invented this magnetic configuration, which uses helical or non-planar toroidal coils to produce a magnetic bottle, which, when properly designed, resulted in good confinement of the plasma. Avoiding the massive plasma current, which the tokamak needs to provide a twisty field, is a huge advantage.

Unfortunately, the Princeton Plasma Physics Laboratory (PPPL) did not succeed with the approach and, in the late 1960s, gave it up for the tokamak. The stellarator flag was then carried by Uo's group in Japan, by Russian laboratories, and by the German laboratory at Garching.

Uo-san invented the heliotron variant of the stellarator, which uses a single helical coil wrapped multiple times around the vacuum torus—19 times in the case of his final experiment, Heliotron-E. My office mate on JET, Denis Marty, and his colleague Charles Gourdon, came up with a similar device, the torsatron, which used two coils wrapped around the torus (Figure 12.1).

Sometime in 1965, I met Koji Uo for the first time when he joined the Tarantula group at Culham for a year. Before arriving in England, he and his family had been living outside of Japan, first for a year at PPPL, then for a year at Garching. Their odyssey had resulted in their children being multilingual. The first child was born in Japan and spoke Japanese and English. The second, who learned to speak in the United States and Germany, spoke mainly English and German. The youngest spoke mainly English.

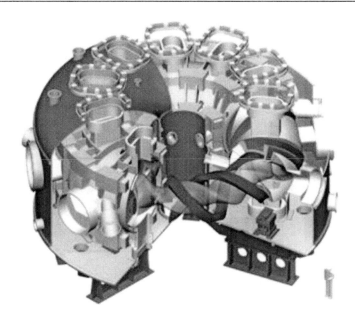

Figure 12.1 The most recent version of Koji Uo's vision, a cutaway of the Large Helical Device (LHD) showing the helical coils.
Source: courtesy of the National Institute for Fusion Science, Japan.

Uo-san claimed that he spoke both English and German. Maybe, but his English was hard to understand. Consequently, when he announced that he and his wife had planned a trip to Spain, leaving the children with a nanny, we were concerned, not only for the children, but also because the couple was not going with a tour group. We wondered how Uo-san would deal with the Spanish language.

The Monday after the trip was over, and during a morning coffee break, we were surprised and relieved when the small figure of Uo-san came through the door.

After he joined us with his coffee, Jim Paul asked, "How was your trip?"

"Excellent," Uo-san replied.

"How did you manage with the language?" someone else asked.

"No problem. I speak Spanish."

"You do?" we all echoed.

"Yes. To Japanese, all European languages are the same."

Then we understood why we had difficulty following his English. It was an amalgam of English and German, and soon to be Span-ger-lish.

It is well known that the Japanese have a hard time distinguishing the sounds of the letters *L* and *R*. For example, when a friend was on a sightseeing tour in Tokyo, the tour guide announced proudly that they were now approaching the "Parace of the Clown Plince."

Of course, the Japanese are aware of this difficulty. Some years ago, a Japanese laboratory proposed an experiment to include a tiny amount of tritium, a limit imposed because of their strict standards on radiation exposure. It was a very modest device compared to ones that were being developed to achieve ignition in a deuterium-tritium (D-T) plasma in the United States and Europe. The Japanese named the experiment the R-Project. I was curious what the "R" stood for and asked Hideo Ikegami, a colleague whom I knew from my Elmo Bumpy Torus (EBT) days, for an explanation.

With a straight face, Ikegami-san replied, "The *R* stands for *Learner.*"

The story reminds me that during the late 1970s and early 1980s, Jerry Kulcinski, from the University of Wisconsin, was a member of the multinational INTOR team in Vienna (the forerunner to ITER), who were designing the next large tokamak. As is typical in such ventures, the team was split into groups to tackle specific components of the proposed experiment. One or more Americans, Europeans, Japanese, and Russians were assigned to each group. The chairmanship of the groups was divided among the participating countries. In Jerry's case, the chairman was Japanese.

After many weeks of presentations and discussions, the group was ready to write its report. Jerry suggested to the chair that, since English was his native language, it might be simpler if he wrote the first draft of the report. The chairman appeared surprised but agreed.

Jerry worked hard over the weekend to prepare the draft, and he gave it to the chair late on Sunday.

On Monday, the group met. The chair rose and said, "We all owe Professor Kulcinski a debt of gratitude for spending his weekend in writing this draft. Thank you, Professor. However, I do feel it necessary to comment that in the third paragraph of the second page, there is a totally incorrect use of the subjunctive."

Of course, most English-speaking people these days have little idea what the subjunctive is. (You could have included me before this story made me think about it.)

The Advanced Toroidal Facility (ATF)

Returning to the late 1970s, results were coming in from the ISX-B tokamak program. Our main effort was in testing plasma pressure (beta) limits. At first, the pressure rose linearly as we increased the neutral beam power into the plasma. But as we went to higher powers, the pressure went up more slowly. We had discovered that the confinement of plasma energy decreased with the heating power—what became known as the low- or L-mode of operation.

Our results, quantified by Masanori Murakami and Dave Swain in 1981, when extrapolated to the large tokamaks coming online—JET, JT-60, and the Tokamak Fusion Test Reactor (TFTR)—did not auger well for their performance. Sponsors do not like bad news, and the initial reaction was that we must be doing something wrong. But when data came in from other tokamaks, it became clear that degradation of performance with increasing power was a fact of life.

At a meeting in Varenna, Italy, in 1982, the situation began to look better. A German team announced that they had found a regime of higher confinement on the ASDEX tokamak—the high- or H-mode. As they raised the heating power, the energy confinement time first dropped, but then, spontaneously, it jumped to about twice that level. From then on, the degradation with increasing power continued, but extrapolation to the larger tokamaks was now encouraging, in that it appeared they would be able to meet their goals.

Hints of improved confinement had been seen in ISX-B and other tokamaks, but these experiments all had their plasmas resting directly on a material limiter. The feature of ASDEX that had allowed them clear access to the H-mode was their magnetic divertor, which isolated the core plasma from a material boundary. Subsequently, both JET and JT-60 installed similar divertors, while the TFTR experimentalists developed an alternative way of operating the plasma to make a respectable amount of fusion power.

Shortly after this breakthrough, it became clear that the Office of Fusion Energy (OFE) would not continue supporting ISX-B. The belief was that the larger tokamak experiments would be more relevant. We proposed ISX-C, a steady-state tokamak that would study divertors and plasma-material interaction problems. This experiment fitted well with the strength of the Oak Ridge National Laboratory (ORNL) in materials research and plasma technologies. In addition, we were one of a very few fusion laboratories that had tens of megawatts of steady-state power supplies and water cooling towers available. With the help of Murray Rosenthal, we succeeded in getting construction funding identified in the fusion budget. To help bridge the gap between the ending of OFE funding for ISX-B and the start-up of ISX-C, I managed to obtain a contract to test a beryllium limiter for JET. Beryllium dust is highly toxic, and this was a test that only an experiment going out of business would consider doing.

Unfortunately, the best-laid plans ... I suspect that it was some of our "friends" in the fusion program who persuaded OFE that ISX-C was not needed, because their experiments would do the needed research. OFE told us to cease work on the design, and 25 years later, the steady-state work that we proposed has *still* not been done!

Fortunately, Murray Rosenthal was able to persuade OFE that since the funds were identified for ORNL in the congressional budget, we should be given the opportunity to come up with another proposal—of course, one that would have to pass a strict review.

We assembled our key experimental and theoretical staff and brainstormed. Clearly, another tokamak variant wouldn't sell. A major reversed field pinch was in the Los Alamos program, as was a compact torus program. Eliminating those options led us to the stellarator, which was being kept alive in a modest but effective experimental program at the University of Wisconsin.

Extensive theoretical studies led by Jim Lyon, with Ben Carreras, Ron Fowler, Geoff Harris, Steve Hirshman, Tom Jernigan, Jim Rome, and Uncle Tom Cobley (an allusion to an old English song), and all arrived at a torsatron as the stellarator of choice for the Advanced Toroidal Facility (ATF). The particular feature that convinced us to make this choice was that in this configuration, it would be possible to have a relatively smaller aspect ratio than most stellarators tested up to that time—a ratio of 7 rather than the more common 15 to 20—and more relevant for a reactor. Our theoreticians also calculated the possibility of a decent beta around 5 percent.

The success of Koji Uo's heliotron program and major advances, both theoretical and experimental in the program at Garching, were also among the reasons why we chose the torsatron/heliotron configuration.

＊ ＊ ＊

My favorite memory of Uo-san concerns him sitting cross-legged on a dining-room chair at a party he and his wife gave when they were in England. He had imbibed quite a lot of scotch and was unwisely lecturing us on various topics about war. When we countered his arguments, he told the following story to illustrate that our approach to life was seriously flawed.

"The problem with you Westerners is that you are too logical," he said. "Your kind of logic."

"What do you mean?" I asked.

"I will tell you the story of the young American who came to Japan in the 19th century," Uo-san said, taking a sip of his whiskey.

On arrival in the country, the student went to the university in Kyoto and spoke to a professor, explaining that he wished to understand Japanese culture.

"Then you should study Japanese archery," the professor said.

When the young man arrived at the archery school, he saw various bows and arrows. He picked one up and was about to take an arrow when a voice said, "What are you doing?"

"I'm here to learn archery," he replied.

"Not like that," said the instructor. "First, you must achieve the correct attitude. Come back in six months after you have read about archery and considered your approach."

Six months later, the American returned. Satisfied that he had completed the initial phase of his studies, the instructor allowed him to pick up a bow and string an arrow. The instructor then placed a hand on the young man's right shoulder and said, "What will you do now?"

The young man responded by pulling back the bow string and arrow.

"Stop," said the instructor, withdrawing his hand. "Your muscles tensed. You were using force."

"But—"

"Force is not necessary." The instructor took the bow and arrow. "Now, place your hand on my right shoulder."

The student obeyed and, as the string went back, he felt no tension in the instructor's muscles.

"Now go away and consider what I have shown you," said the instructor. "If you feel ready for the next stage, come back in six months."

When the American returned, the instructor gave him the bow and arrow, placed his hand on the young man's shoulder, and said, "Show me that you have learned and are ready for the next stage."

This time, when the bow string went back, the student's muscles were relaxed.

"Very good," said the instructor. "Now put the arrow in the target."

The American pointed the arrow at the target.

"Stop!" said the instructor. "Have you learned nothing? It is not necessary to point the arrow at the target. Go away, and when you have the right attitude ..."

As the saga continued, I wondered whether Uo-san, a physicist, was serious. Later, I concluded he was.

* * *

My last encounter with Professor Koji Uo was on the occasion of his retirement. By this time, ATF had started operation and was producing interesting data that complemented his work on Heliotron-E and the major advances in the German program. I went to Tokyo for a meeting. Prior to going, we at ORNL, and our OFE sponsors as well, decided that it would be a nice gesture to give Uo-san a plaque commemorating his distinguished career. We had congratulatory words and a schematic of the Heliotron-E coils etched on a steel plate that was mounted on polished wood.

Our actions were truly out of respect for the man. Although Uo-san could be rude and must have been difficult to work for, he had developed a superior program and had attracted a number of the brightest Japanese fusion researchers to his team. He was surely one of a kind, and the fusion program owes him a lot.

On a weekend, John Willis (from OFE) and I went by train from Tokyo to Kyoto, and took Uo-san to a hotel near the station for lunch. We presented him with the plaque and told him how valuable his contributions and collaboration had been to the United States and ORNL. Then, Uo-san spent most of the lunch telling us how stupid ATF was—particularly, the choice of material for the vacuum vessel and having helical coils with an even number of turns (unlike his heliotron).

After we said goodbye and were making our way to the station for the two-hour train ride back to Tokyo, John Willis turned to me and said, "Can you explain to me why we did this?"

I told John the story about Japanese archery and informed him that Koji Uo had been captain of Japanese archery during his university days (Figure 12.2).

I don't believe that Uo-san was correct about us having chosen the wrong material for the ATF vacuum vessel. But he was correct that we would have problems. We made the mistake of constructing this complicated component in one piece, by welding plates onto a skeletal structure that defined the inside dimensions of the torus, with its two grooves into which the helical coils would fit. After the vacuum vessel was completed, the skeletal structure was cut out. Unfortunately, we did not allow sufficiently for built-in stresses, which resulted from the shrinkage that can occur during welding.

When we tested the vacuum vessel for accuracy, it became obvious that it would not fit over the lower helical coil segments. The solution was to cut slots, stretch the vessel, and re-weld. In the end, the coils fitted but suffered leaks from trapped volumes in the metal. The only way that we could get adequate vacuum for plasma experiments was to spray titanium onto the inner wall periodically—as welcome a solution as painting over the rust on a ship's hull.

Figure 12.2 Test assembly of the top segments of the helical coils.
Source: Courtesy of the Oak Ridge National Laboratory.

On the positive side, the Japanese at Nagoya took one of our lower-aspect-ratio torsatron designs for their Compact Helical System (CHS) experiment. Results from ATF and CHS provided the data to support the later construction of the massive LHD at a new site near Nagoya. This billion-dollar facility with its superconducting coils has met all expectations in terms of plasma confinement and beta (more than 5 percent to date). Along with the equally large WVII-X stellarator at Greifswald in Germany, LHD is paving the way toward the stellarator approach to a power reactor.

13 Fusion's Prospects

At a conference in Grenoble in the early 1980s, the dinner was held in a chateau that was a guesthouse for the Commissariat â l'Énergie Atomique (CEA). In an after-dinner speech, one of the French organizers told us (in French) a story about the chateau's ghost. An American, who was working at the Garching laboratory in Germany, translated the story into English—a strange decision on his part, I thought. I was sitting with Masaji Yoshikawa, head of the Japanese JT-60 experiment, and one of a number of Japanese at the meeting. I suggested that Yoshikawa-san translate the story into Japanese, write it down phonetically, and I would relate it to the group. He did so and, fortified by plenty of wine, I read his note.

"Was I understandable?" I asked.

"More or less," Yoshikawa-san replied. "Do you know what you told them?"

"About the ghost," I replied.

"Are you sure?"

"You mean I didn't?"

"You told a dirty joke," Yoshikawa-san toasted me with his wineglass.

"You're kidding."

"You'll never know, will you?" said Yoshikawa-san, looking inscrutable.

Fusion research has always suffered from ill-advised statements by people who don't know what they're talking about: *You're stuck in the past. You need to think outside the box. My approach will be commercial in 10 years.* Sometimes, it's a reflection of true enthusiasm, sometimes ignorance. Too often, the people who talk about thinking outside the box don't know what's *inside* it. And sometimes people who in another era would have tried to sell you the Brooklyn Bridge are trying to con funding out of the government or private investors.

Sadly, it isn't difficult to persuade sponsors, and particularly nonscientists, that the researchers are stick-in-the-mud, over-the-hill people who are unwilling to accept innovation. To sponsors, the grass is often greener outside the box. A consequence of this belief is that flights of fancy—cold fusion, bubble fusion, and so on—get an unwarranted level of support. The result is to slow down the ongoing program and make it more difficult to get closure on the important problems that *are* being tackled. Success usually comes from reasoned persistence—different from pigheadedness—and flitting from one approach to another doesn't work well.

On rare occasions, the people who believe that they know enough to make dramatic changes to the fusion energy research program are correct. When this occurs,

Fun in Fusion Research. DOI: http://dx.doi.org/10.1016/B978-0-12-407793-5.00013-3

I am reminded of the statement attributed to Alfonso the Wise, King of Castille (1221–1284):

Had I been present at the Creation,
I would have given some useful hints
For the better ordering of the Universe.

But in order to make a change, it is generally necessary to stop part of the research program. It is hard to know when to pull the plug on an approach, and certainly aficionados will battle mightily to do that one final test that will convince the funding agency to continue its support. Nevertheless, over the years, the fusion program has faced up to closing down unsuccessful lines of research, e.g., the simple pinch, the simple mirror, the EBT. But it is important to not give up too easily. Some lines were dropped in haste in the United States—most notably the stellarator.

The Dinosaur Chart

I created the dinosaur chart after being informed by a congressional staffer that we never wanted to close out any line of research. Curious about what had actually happened in the U.S. program, I obtained historic funding data from 1966 to 1988 on the various magnetic confinement concepts from the OFE. I soon realized that the funding profiles, when plotted in a particular way, looked very much like the plots of the abundant number of ancient species obtained from fossil remains. I wondered: What if, far in the future, archaeologists were digging in the fusion laboratories of the past. They would find strangely shaped vacuum vessels and copper and superconducting coil fragments. From these pieces, they would be able to construct an evolutionary map—the fusion dinosaur chart.

The results of my research are shown in Figure 13.1, which has on the ordinate the percentages of the experimental budget, given by year (1966–1993), to each of the magnetic confinement approaches—mirror, multipoles/Astron, EBT, pinch, stellarator, and tokamak. The burning plasma refers to the deuterium–tritium (D–T) part of the TFTR program. A "dinosaur chart" would show the same kind of information for ammonoids, crinoids, dinosauri, etc., in the Cretaceous and Triassic ages in terms of number of fossils found or estimated.

Now to the subject of people talking nonsense. The best example of overly aggressive speakers being put in their place occurred at a meeting to discuss potential D–T burning successors to Princeton's TFTR. Prof. David Rose of MIT, a highly respected senior scientist, was an attendee. Along with many of us, he listened with increasing irritation to the absurd claims of participants, mainly from the private sector (though, sadly, some came from the fusion community). The private-sector people said, in so many words, "You academics from universities and labs are holding us back. Give us the money. We'll build a fusion reactor."

Finally, David Rose raised his hand and, when there was silence, said in his wonderfully sonorous manner of speaking, "Gentlemen, your claims remind me of a scene from Shakespeare in which Falstaff is telling his friends about all the

U.S. EXPERIMENTAL BUDGETS AS A PERCENTAGE
OF THE TOTAL

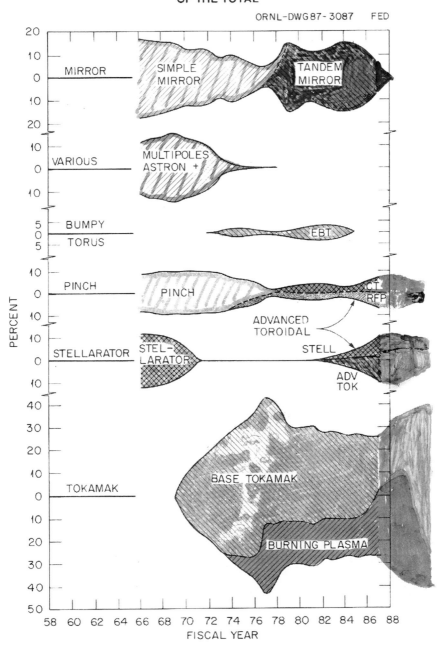

Figure 13.1 The dinosaur chart.
Source: Courtesy of the ORNL.

wonderful things he and they will do. After a while, one of his friends says, 'But Falstaff, how will we do them?'

Falstaff replies, 'We will summon spirits from the deep to aid us.' A long silence ensues before his friend says, 'But what if they don't come?' "

This comment caused most of the unwise participants to temper their claims; however, one aerospace engineer argued passionately, to show that the industry's capabilities were being underestimated, "But we went to the moon."

I managed to get my two cents in. "Yes," I replied. "But you knew where it was."

The reality was that the program was still a decade or more away from having established an adequate database to design intelligently a major, steady-state, D–T burning tokamak. In fact, it wouldn't be until the 1990s before the large tokamaks, TFTR, JET, and JT-60U achieved many of their goals.

The Art of Stating Goals—JET and TFTR

Why do scientists overstate the rate at which they can achieve their program's goals?

The answer to the question in the title is simple; it's because the funding agencies don't want to hear the reality. This point is well illustrated by the actual schedule of the large D–T burning tokamaks, JET, and TFTR.

TFTR had its first plasma in December 1982. JET, whose construction had been delayed by European bickering (as explained earlier), followed in June 1983. At the time, I heard that a key DOE official was amazed and annoyed to hear that TFTR did not plan to start its D–T phase of operations within a year of start-up. The reality of fusion experiments is that, unless they are a clone of an existing experiment, they are breaking new ground, and it takes time and experimental skill to make them hum.

The highest plasma current in a tokamak to that time (Princeton's PLT tokamak, which was the first to achieve muti-kilovolt electron temperatures) had been less than 1 MA, and TFTR was designed for 2.5 MA and JET for more than 3 MA (finally, it operated up to 6 MA).

In addition, when ASDEX demonstrated the H-mode of good confinement, it was with a magnetic divertor that isolated the core plasma from the plasma edge. Neither TFTR nor JET had a divertor, although a poloidal divertor similar to that in ASDEX had been designed for JET in the mid-1970s by Peter Noll of the JET team. The configuration of TFTR, with its circular plasma, did not lend itself to a subsequent installation of a divertor.

In Japan, the JAERI team upgraded the original JT-60 to JT-60U to include a poloidal divertor. Later, JET installed one with the coils inside the vacuum vessel.

As an aside, I had the privilege of celebrating the start-up of JT-60. The Japanese have a delightful custom for such events—the Daruma doll. I was told that Daruma was a monk who spent years sitting in front of a wall, staring at it

Figure 13.2 In the back row (from left to right), Masaji Yoshikawa, Kenro Miyamoto, Anne Davies, Harold Furth, Shigeru Mori, Ken Tomabechi, and Christine Ludescher.

until he went blind. The dolls honor his memory (along with Japanese archery—there is a message here). At the start of a journey—or construction project—a doll is purchased and team members ink in the left eye, a dot at a time, using a calligraphic pen and black ink. At the completion of construction, team leaders and honored guests ink in the other eye.

Anne Davies, director of the OFE, Harold Furth, director of the PPPL, Christine Ludescher, his wife, and I were honored to participate in the ceremony (Figure 13.2).

What's in a *Q*?

All the large tokamaks performed very well, meeting their important technical goals, including operating successfully with D–T in JET and TFTR: nevertheless, an important lesson in how to define your program goals came out of the TFTR experience.

The experimental goals for JET were kept vague, deliberately and wisely, at the insistence of Paul Rebut, the director, and Donato Palumbo, the head of the Euratom Fusion Programme. The stated goal was as follows: "The essential objective of JET is to obtain and study a plasma in conditions and dimensions approaching those needed in a thermonuclear reactor." No claim was made as to what the maximum Q achieved would be ($Q =$ fusion power produced/heating power to the plasma).

The goal was to have kilovolt temperatures (already achieved in a smaller experiment) and to study alpha production and heating.

I helped to put together a publicity document for JET (JET R-7) that showed a picture of the operational space (density $\times$ confinement time versus temperature) that JET might be able to access.

In contrast, while Princeton had written similar scientific objectives to JET, they latched onto the Q-parameter as a way of selling TFTR (I think this shows the difference between the U.S. and European approaches). Specifically, when senior managers talked about TFTR, they emphasized reaching $Q = 1$, which was called *breakeven.* This kind of parameter is easy for nonscientists—such as the U.S. Congress, the press, and the public—to understand. But as soon as TFTR operated, Princeton staff were asked when they'd achieve $Q = 1$. In fact, it was not until November 1993, 10 years after start-up, that D$-$T was introduced into TFTR, and during the next 4 years, PPPL mounted an aggressive campaign to maximize the fusion output, ultimately producing 10 MW ($Q \sim 0.3$) of fusion power for a fraction of a second. More importantly, they used an impressive array of diagnostics combined with theory and computer modeling to study the physics of alpha-particle behavior. Unfortunately, they were still being held to their simple goal.

Around this time, the problem was compounded during one of the yearly visits of senior fusion managers (including me from the ORNL) to "the Hill." We were in the offices of the chair of the U.S. Senate's energy committee, meeting with an aide.

"How's progress on reaching Q of 1?" the aide asked the Princeton delegation.

"Going well," Harold Furth replied, stroking his beard. "We have some wonderful data on the alphas and it seems in good agreement with theory."

"Yes, yes, but what about Q?"

"I think we'll get there, but it takes time," Harold replied calmly.

At that point, the other Princeton attendee decided to intervene. "It's going very well," he said. "Forget about Q of 1. I hear that people think we'll get Q of 2."

Horrified by this rash statement, most of us looked down at the floor.

"Great," said the aide with a tight smile. "I'll write your new goal into the appropriations bill."

I don't think he did, but the bill did contain language that held Princeton's feet to the fire.

In the end, they achieved Q of around 0.3, and despite their superb scientific studies, TFTR was viewed by many nonscientists as having failed.

As for JET, they did a quick test with 10% tritium in November 1991, achieving about 1 MW of fusion power. Subsequently, when they went to 50% tritium, they achieved 15 MW for a second at $Q = 0.7$. Of great importance, both facilities showed that it was possible to handle significant quantities of tritium safely.

The Shiva Winner Altruistic Trust

What the program needs, even today, to discourage scientists from making unrealistic claims are more humorists like Tudor Johnston, a professor at the Université de Québec. Irritated by the claims being made in 1977 by Livermore scientists for their soon-to-operate inertial fusion Shiva laser system, Tudor submitted a poster to the annual meeting of the APS on plasma physics, held from October 31–November 4, 1978. His poster would reveal the results of a winner-take-all pool of estimates of the maximum yield of neutrons achieved from compressing a target on Shiva within a short time of operation. The competition was organized by the so-called Shiva Winner Altruistic Trust (SWAT), with Tudor Johnston as spokesman and trustee. February 1 was the deadline for a "member" to submit a bet, accompanied by a $1 bill inscribed with the member's name, organization, and date. All contributions would go to the winner.

On January 31, 1978, the deadline was extended to March 1. In the end, 79 people responded. Early estimates from Livermore ranged from 32 billion to 460 billion fusion neutrons per implosion with an average of 540 billion neutrons.

Later, encouraged by Tudor, other Livermore scientists made estimates. They ranged from 83 billion to 500 billion, with an average of 91 billion neutrons. The target designers were, in general, more optimistic than the theoreticians.

Estimates from other institutions included zero, and then ranged upward to 1.23456789 billion to 1.2 million billion, which averaged, ignoring the extremes, 65 billion.

The actual achievement was 27 billion. The winner was Don Slater of KMS Fusion, who guessed 27.5 billion neutrons; he won $79. He was closely followed by Jack Wilson of the University of Rochester's Laboratory for Laser Energetics, who volunteered the speed of light in centimeters per second—29.97929 billion (Table 13.1). Jack Wilson's submission is shown in Figure 13.3.

I believe that Shiva did eventually meet its goal, but it took a lot longer than anticipated; an important message is here for the National Ignition Facility (NIF) ignition campaign.

Interestingly, another varied estimate of size occurred when I was touring the NIF facility a few years ago. This system of 192 powerful lasers looks big enough to fill a Walmart store (see Figure 17.10). The laser beams converge to focus at the millimeter scale onto a target, delivering up to 2 MJ of energy. It is expected that the device will ignite a D–T-filled fusion target and produce a gain in energy. My tour occurred a month after the official opening of the facility by numerous dignitaries, including California governor Arnold Schwarzenegger. I had once heard a

Table 13.1 Neutron estimates taken from the Analysis of SWAT Predictions

Actual	27 billion neutrons per shot	27
Winner	27.5 billion	27.5
Livermore	83 billion to 500 billion	91 billion average
Non-Livermore institutions	Zero to 1.2 million billion	65 billion average, ignoring extremes

Source: Courtesy of Tudor Johnston.

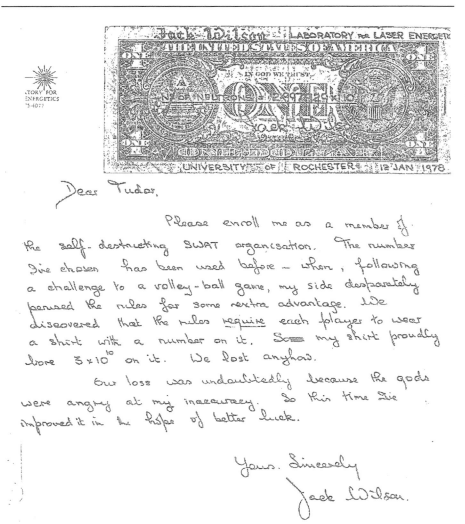

Figure 13.3 Jack Wilson's bid.
Source: Courtesy of Tudor Johnston.

debate about the governor's height, so when we were touring, I asked my guide whether she had met him and, if so, how tall he was.

She replied, "I'm five-ten, and he's about my height."

Later, I came across Ed Moses, head of this area of research at LLNL. I told him what I had heard and asked him his opinion. "No way," he said. "Schwarzenegger's well over six foot."

The next day, puzzled by this disagreement from people who had stood next to the man, I asked a colleague if he would Google Schwarzenegger's height. He looked surprised but kindly performed the task. He was even more surprised to find a host of sites. The one that appealed to me most was the one that stated, "Estimates of Arnold Schwarzenegger's height range from 5 foot 6 inches to 6 foot 5 inches." Maybe there's more than one of him?

The next time I bumped into Tudor was at an APS meeting in Los Angeles in the early 1980s. I had been sitting on my hotel bed on the seventh floor trying to gather the energy to go down to the conference. There would be 1500 attendees, 12 parallel sessions, and an infinity of dull talks and posters. As I dragged myself out of the room, I was praying for something to liven up the day. I rounded the corner, and I saw someone I knew standing in front of the seventh-floor elevator—Tudor.

"Tudor, nice to see you again," I said enthusiastically.

Tudor turned around and grasped my hand. "The pleasure's mine. I'm not having much fun. I've been standing here for 5 minutes."

At that moment, the elevator arrived. The doors opened, revealing a sardinelike mass of uncomfortable-looking riders. No ride for us.

"I am so glad you could come to the meeting," said Tudor.

Startled faces peered out at us as he continued. "We have a busy agenda so we should get on with the first item. It is—"

The doors closed to the sound of nervous giggles. Tudor turned to me. "I love to do that," he said.

14 Fear and Flying

Harold Furth and Mel Gottlieb from PPPL, Chuan Liu from the University of Maryland, and Marshall Rosenbluth, then at the University of Texas, made one of the first visits to China after the last of Mao's disastrous "great leaps forward." All were quite large men in their different ways: tall, wide, sturdy, or all three. Westerners were a rare sight in China in the 1980s, and they attracted crowds everywhere they went. In the market in Hefei, they noticed that two tiny, old Chinese ladies were following them closely, jabbering excitedly.

"What are they saying?" Harold asked.

Chuan Liu grinned and pointed to the smaller of the two women who was staring up at them. "She said, 'Look at the size of them! If that's what the West is like, we'll never catch up.' "

Fear

The ORNL is one of the premier laboratories in the world for materials research and development (e.g., cladding for reactor fuel, whisker-strengthened ceramics, inter-metallics, improved steels, and new carbon forms). Nuclear reactors and an array of other facilities support this research and other studies as well, including a facility to handle tritium back in the 1980s.

In the early 1980s, I led a tour of a group of Japanese fusion scientists around the ORNL "Tritium Facility" and the High-Flux Isotope Reactor (HFIR). One of the scientists ran the Japanese tritium facility for fusion research in a doubly contained building with subatmospheric pressure to ensure no leaks. His total tritium inventory was something like a tenth of a gram—1000 Curies (a Curie is the quantity of a radioactive isotope—such as tritium—undergoing 37,000 million disintegrations per second).

Our facility consisted of a hut in the woods. The tritium was contained in three standard shipping cylinders, which were attached to a stainless steel vacuum system. The tritium was stored on uranium beds within the cylinders. The whole apparatus was mounted under a transparent hood. A more noticeable feature was the gale of air that swept past the cylinders and went to a vent in the roof; this function was designed to minimize exposure to personnel in the event of a leak.

A technician, wearing a lab coat and cowboy boots with one trouser cuff in and one out, started to describe the facility. The Japanese scientist took out a notepad and pen.

Fun in Fusion Research. DOI: http://dx.doi.org/10.1016/B978-0-12-407793-5.00014-5

"We have 240,000 Curies here," he said.

The scientist, talking to himself, started to write. "Two-a-hundred-and ..."

"Yes, sir, 240,000."

"Two hundred and forty ...." The scientist started to add a zero. "Two a hundred and forty thou ...." He finished the third zero and looked up. "Where you got this two hundred and forty thousand Curies?"

The technician took him by the hand and, reaching under the hood, he placed it on one of the cylinders. "Right here, sir," he said. "80,000 Curies."

The scientist sprang back, looking at his hand in horror. He retreated to the far side of the hut.

We continued on to HFIR. As soon as we arrived, the scientist told me he would like to wash his hands. As he went into the restroom, I said to his colleagues that I assumed he needed a toilet.

"No," one of them replied. "He wants to wash his hands."

I glanced around the door and saw the scientist, with his jacket off and his sleeves rolled up, frantically washing his hands and forearms.

HFIR is a so-called swimming pool reactor. The reactor operates under tens of feet of water that provide complete shielding from the neutrons and gamma rays. The highly enriched uranium cores only last some 20–30 days and are stored in a separate part of the pool for their radioactivity to decay. They glow with the eerily beautiful blue light of Cherenkov radiation. To the amusement of his colleagues, our friend refused to go and look over the railing. He'd had enough excitement for one day.

What's in a Name?

While people's names are interesting, the reason people change them is more fascinating. I recollect reading an obituary from the *Lamp*, the journal of the Liverpool and Merseyside Polytechnic, quoting the *Liverpool Free Press*. The obituary concerned Kanso Yoshida, cousin of Emperor Hirohito of Japan. Yoshida had died in Liverpool at the age of 78. It stated that since coming to Liverpool in 1912, he had lived incognito under the name Paddy Murphy. I later heard the opinion that students had invented the story. At the time, it seemed a likely explanation. My own experiences since then have made me reconsider that view.

In 1985, I became a U.S. citizen. The ceremony took place in the federal courthouse in Knoxville, Tennessee. I was somewhat wary about taking this step, remembering stories from an English man and woman who had undergone, separately, the following similar experiences in Texas. The main comments of the judges appeared to have been developed for illegal immigrants from Central America. Both judges talked about the virtues of the Texas prison system and how everybody was entitled to a fair trial. In the case of the woman, the judge went on to add, "And now you are free."

She stormed out of the courthouse. As she did, she retorted angrily, "I am British, and I have always been free. I had hoped to be free here. Your comments make it sound uncertain. I will not proceed further with this nonsense."

My experience in 1985 in the federal courthouse in Knoxville was much more pleasant. The judge gave a thoughtful and relevant speech, reminding us of our obligations when we became citizens and telling us to be clear about whose side we would be on in the event of a conflict with our previous country. Someone then read out the list of new citizens. Until that point, I had not realized that becoming a citizen was an opportunity to change one's name. The first to do so was a gentleman, possibly from Thailand, called something like Boong Bin Gong Yoon Gang, who now wished to be known as Boo Gang. A sensible move, I thought.

Subsequently, I learned that Mei Mei Hwang would become Michelle Hwang. Others followed with similarly simple, easy-to-understand name changes—although it became hard for all of us not to chuckle at some of the preferences.

But even the judge could not contain his laughter when a sweet-looking Vietnamese girl stood and the clerk announced that Hua Lo Duc would now be known as Heather McTavish.

Flying to D.C.

After I became the director of ORNL's Fusion Energy Division and Program in 1988, I traveled to Washington, D.C., a lot. I had the same reason as most people, I guess—that's where the money is. On one occasion, thunderstorms between Knoxville and Atlanta made the connecting flight really bumpy, and we were even hit by lightning—scary as hell! When we landed in Atlanta, I waited in my seat as passengers ahead of me removed their bags from the overhead bins. As a man retrieved his bag, a laptop computer that some idiot hadn't secured fell out and landed on an old lady's head. She was hurt and bleeding, so the flight attendant called for help, and medics carried her off the plane.

I hadn't realized that my colleague Bob had been sitting behind me, but he caught up with me at the gate for the D.C. flight. I had worked with Bob for about 10 years on ITER topics. He was an expert on neutronics—calculating radiation shielding and so on—a great guy and one of the most charming men I've ever met. He had a lovely wife and daughter. It sounds strange to say this about another man, but Bob had the greatest smile—as they say, it could light up a room. This is my memory of our conversation:

"Hi, Bob. Did you see that computer landing on the poor old lady?"

"Yeah, I hope she's okay." Bob paused. "You know it's weird but a similar thing happened a year ago ... on the next leg to D.C."

"You're kidding."

"No. We were in a 757. You know, three-across seating on each side. I was in an aisle seat near the back. We hit the worst clear-air turbulence I've ever experienced. I thought we'd crash. The captain told us to buckle down as tight as we could because it would be another 10 minutes before he could get us out of it."

"At least you didn't get lightning."

"Small comfort, I'm afraid. About a minute later, the bin above me opened up and, as the plane lurched, a large case flew out and hit the man sitting one row up on the other side of the aisle."

"Good God!"

"The man was hurt. A flight attendant saw what happened, and with great determination, she struggled down this bucking airplane to see if she could do anything to help him."

Bob stopped and flashed that joyous smile. "The most incredible thing happened. Just as she approached his seat, the plane hit an air pocket. She shot up in the air and came down head first over the guy in the seat in front of me. Now wait for it: Head first down and between my legs and stuck."

"No way."

"I tried to lift her up, but I was buckled in so tight I couldn't exert any leverage. We were stuck in this weird position until the turbulence stopped and my neighbor helped me lift her up. Helluva girl ... she immediately went to help the injured passenger. When the other flight attendant arrived, the heroic one went to tell the captain what had happened."

"Did it all work out okay?"

Yes, in the end yes. But there's more. A short time later, she came back down the aisle, looking embarrassed. "I don't know quite what to say, sir. Nothing like that ever happened to me before, and it isn't in our training."

I hadn't thought about what to say either and, without thinking, I smiled and blurted out, "I think I love you." She turned bright red and fled back to her seat.

"You devil," I said, laughing.

Bob shrugged. "It just came out that way." His look was beatific.

On another occasion, I took the direct flight back to Knoxville from D.C. When our commuter flight was called, I joined the other passengers on the bus—apparently, a recycled, rental car courtesy vehicle, with seats all around the sides—that would take us to our plane. The commuter airline was headquartered in Florida, and each plane was named after a Florida city: Miami, Tampa, St. Petersburg, Daytona, and so on. Our flight, scheduled for 7:10 p.m., had been delayed by bad weather to the west, and it was now 8:00 p.m. The rain started as we drove around in the dark looking for our plane. I remember thinking, *Thank God, there it is.* The name signed in script on the front was Orlando. The bus stopped, and we rushed to our seats.

A bright-faced flight attendant greeted us with, "Welcome aboard. We'll be leaving soon for Charleston."

A chorus of voices shouted variants of, "What?"

The flight attendant looked worried. She opened the door to the cockpit, and we heard a brief, muffled conversation. She came back into the cabin, shaking her head. "Sorry folks, this plane's going to Charleston, West Virginia. The bus will come back for you."

We boarded again in silence, getting quite wet in the process. I looked out at the dark and rain-drenched tarmac, as the driver completed a crackly discussion with his microphone. He turned to face us and said, "I hate to tell you this, but your equipment's circling Cincinnati." Seeing most of the men look down, he added. "Don't worry. I was referring to your plane. They're trying to get another one. I've been told to park near the terminal."

The bus circled back to some dark, drab area, away from our terminal. When the driver opened the door, a mist of rain invaded the front of the bus. "Please wait here, I'm going to the office to find out what's happening. I'll leave the engine and interior lights on," he said as he left.

After a moment of silence, in which we saw occasional flashes of lightning, an elderly man said, "I know what's happening. We're in the Twilight Zone—condemned to sit in this bus, parked or driving around this godforsaken airport in the dark and rain . . . forever."

We all agreed with him.

Interesting Sides of Al Gore

Sometime around 1990, I attended a fascinating meeting of the National Academy of Sciences and the equivalent Russian Academy. The topics were space and fusion research. On the fusion side, I joined Paul Rutherford of PPPL and others.

One interesting exchange occurred when someone asked a senior Russian about their military budget. The Russian responded, "That is a good question. I can only say that recently, when Mikhail Gorbachev was asked that question, he responded that he didn't know. Does that tell you something?"

Dinner that night was at La Colline, a restaurant near the Capitol. The after-dinner speaker was Senator Al Gore of Tennessee. He gave a fascinating talk on global warming, with a foldout map of the world that showed increasing desertification. Later, I heard that, despite not being a scientist, he read all the important papers on this topic. I shared a taxi back to the hotel with three other attendees, including the son of the distinguished Russian scientist Igor Kurchatov. The son hosted the Russian equivalent of the American TV show *Nova*. I asked what people thought about Gore's talk. Kurchatov replied that it was remarkably good and that he couldn't think of any Russian politician who could have given it.

A few years later, after Gore had become vice president in the Clinton administration, I was attending a fusion meeting at the DOE site in Gaithersburg, Maryland. When the meeting ended at noon, I traveled to downtown Washington to a second meeting in the Old White House. I was among a number of fusion leaders who had been invited to meet with one of Vice President Gore's staff, who wanted an update on the state of fusion research. In fact, I had encountered the staffer when I had attended a congressional investigative committee meeting. I arrived at the Old White House before the others and roamed the halls trying to find our meeting room. It was confusing because office after office was simply called, "Office of the Vice President." Fortunately I bumped into my host, who took me into an office that, yet again, was labeled "Office of the Vice President."

A charming woman ushered me into a waiting room and offered me coffee. After a moment, I noticed that the walls were covered with framed cartoons mocking the vice president. Strange, I thought, that a staffer would so blatantly make fun of the boss, who always appeared to be so serious. It wasn't until I was leaving that it dawned on me that this was actually the office of the vice president. I concluded that Al Gore was very different from the image often portrayed by the media.

Interrupted Talks

At one time, I had considered putting the following stories into a book set in the airport bus I mentioned earlier, as passengers while the time away waiting for a plane that never appears.

In the mid-1990s, I was invited to give a talk on fusion energy at a conference. The main session had five half-hour talks covering energy efficiency, fossil energy, renewables, fusion, and nuclear energy. I was the fourth speaker. The fifth speaker was a well-known scientist. The meeting was held in a convention center and our room, seating a few hundred people, was created by a partition dividing off one end of a much larger hall.

The first talk went well and there was a vigorous question-and-answer session when it ended. A distraction during the second talk was a low hum of music apparently coming from the building's speaker system. During the third talk, this music surged in volume a few times, drowning out the speaker's comments. One of the meeting organizers left to ask for the Muzak system to be turned off. He returned shortly afterward, shaking his head, and said, "It's not Muzak. It's coming from another meeting."

The music wasn't the only problem. Frequent loud asides from the well-known scientist to acolytes on each side of him also interrupted the speakers. I had the misfortune to be sitting close behind this man and could hear clearly his often-inaccurate comments. Furthermore, his questions at the end of each presentation were often put-downs.

Before I could be introduced for my talk, he stood and reached for his viewgraphs, clearly believing that it was his turn to speak. I tapped him on the shoulder and said, "I think I'm next."

He turned, looking irritated, and as the chairman announced my talk, he sat down, muttering, "Whatever."

I managed to finish my talk more or less intact, although the music, now clearly involving a large choir, drowned out a few points with hallelujahs.

Then came the final speaker's talk. The audience clapped politely when he was called to the dais.

"I have what I believe to be a most important proposal to improve—"

"HALLELUJAH!"

The only way we could find out his idea was to read his viewgraphs. At precisely those moments when he wanted to emphasize something, the choir would burst into ecstatic song, their volume rising and falling, seemingly in synch with his attempts to make key points.

After 15 minutes, following a deafening blast of "HALLELUJAHS," he raised his hands pleadingly to the sky and shouted over the music, "My God. What did I do to deserve this?"

The audience knew the answer, and many of us were smiling. There was no question-and-answer period when he finished. He couldn't hear the questions. The chair called the meeting to a close.

As we left the meeting room, the session next door was also emptying, and we discovered the source of the music. An African-American reverend was leaving the hall, resplendent in a purple suit, purple cowboy boots, broad-brimmed purple hat, and copious amounts of gold jewelry. A choir of large African-American ladies, wearing flowing lilac-colored choir robes and ornate floral hats, surrounded him. Trailing them were about 10 people in simple white clothes, with the rest of the congregation close behind. The crescendos of music that had overwhelmed our meeting were the sounds of the faithful in white being saved from their sins— Hallelujah, Hallelujah. I suspect that many of us wished that our last speaker had been among them.

An interruption of a different kind occurred at sea.

Bruno Coppi, a distinguished and inventive professor at MIT, gives enthusiastic talks but sometimes has more wisdom to pass on than will fit in the allotted time. One year, the annual Fusion Power Associates (FPA) meeting was held on a ship that sailed for the day from San Diego to Ensenada, Mexico, and back. On the way back, we all went into the ship's movie theater and listened to presentations. Bruno still had plenty to say about the effects of microinstabilities on magnetized plasmas as his time ran out. The chairman, Steve Dean, president of FPA, didn't say anything to cut him off because Bruno was the last speaker. Strangely, as Bruno continued talking, people wearing Hawaiian shirts and shorts were coming in and taking the empty chairs. All became clear with the announcement over the ship's loudspeaker system: "The movie *Romancing the Stone* will be shown in the movie theater at 5 o'clock."

Bruno gamely lectured the new arrivals about ion temperature gradient instabilities, but the opening movie credits that appeared on the screen made his final viewgraphs with their complicated equations harder to read. Finally, he gave up and packed his remaining material into a small suitcase. There was no time for questions, but I'm sure the tourists had quite a few.

A colleague of mine was on an international committee. One of the other members, a talented European, had command of many European languages and would switch from one to another with ease. The only problem was that he spoke all of the languages in a monotone, with no inflections and, seemingly, no punctuation. When he spoke in English, the common language of the group, his lengthy contributions to the discussion would cause one of the Russians to slump back in his chair and close his eyes until the monologue was over.

Then, on one historic day, someone asked our linguistically able man a question, and he replied, simply, "Yes."

The Russian smiled, sat up in his chair, his eyes alert, watching the man take a sip of water.

The European put the glass down. "Were it not for the other contingency factors that one should inevitably take into account when . . ." the Russian's smile slowly disappeared . . . "considering issues of this magnitude that might affect the whole outcome of . . ." the eyelids closed one after the other . . . "our deliberations and indeed come back to haunt us at some future meeting in which we would be obliged . . ." the Russian sank back into his chair . . . "yet again to investigate the possibility that effects of the kind mentioned could occur in fact and destabilize the whole situation which would . . ." finally asleep.

My final story on this topic is about one way to respond to interruptions.

An ORNL employee gave a series of lectures on behalf of DOE for government officials. His problem was this guy who continually questioned everything he said. For the first day, the lecturer was polite and tried to answer, but this idiot never seemed to listen.

On the second day, the same thing happened. After the first dumb question, he'd had enough.

"Sir," he said, "I have tried for more than one session to deal with your comments and answer your questions. I am having difficulty knowing what to do now. I seem to have two choices. I can continue to try and satisfy you, or I can tell you to stick it in your ear . . . and that's what I am going to do. *Stick it in your ear.*" His audience cheered, and the man never asked another question.

15 The Oscillating Fusion Program

I remained as director of the Fusion Energy Division until 1994. It was a difficult period, with fusion budgets decreasing when inflation was taken into account.

In addition, the program oscillated when a succession of well-intentioned committees changed the priorities. Back in the 1980s, because of uncertainties in the performance of tokamaks, the program had been broadened to include more effort on the so-called alternates—the stellarator, reversed field pinch, and compact tori. ATF was funded as a part of this change.

But, by the end of the 1980s, the tokamaks were doing better, with ion temperatures over 300 million °C, electron temperatures over 10 million °C, and beta up to 6%. Proponents of getting on with fusion energy argued for concentrating on the tokamak. After ATF suffered a serious arc on a helical coil joint, even though the coil had been repaired, it was mothballed in 1991. The reversed field pinch and compact torus research at Los Alamos were also terminated. In universities and the private sector, other compact torus work was also severely curtailed. The decisions were announced at a meeting in DOE–Germantown, Maryland—on my birthday. While I supported the tokamak as the front-runner, I remember commenting that it was a foolish decision by the DOE to increase the tokamak part of the experimental budget from 93% to 97% by destroying the alternates program. Subsequently, ATF was restarted, but then, after a brief operating period, it was shut down for good in 1994.

Once JET and TFTR had begun their experimental programs, the principal debating point in the tokamak area had been what to do after them. During the 1980s, I attended numerous meetings at which we argued about the alternatives:

- An engineering test reactor, whose name evolved from INTOR to ITER.
- A compact burning plasma experiment (proposed originally by Bruno Coppi and his colleagues at MIT), including Bruno's Ignitor, and the Princeton Compact Ignition Tokamak (CIT), which evolved to the Burning Plasma Experiment (BPX) when an advisory panel wisely declared that ignition could not be guaranteed.
- A plain hydrogen experiment to study steady-state, high-powered plasma operation.

I chaired the Ignition Physics Study Group, which was set up to review the database for the CIT. During our deliberations, the committee concluded that a number of areas needed much-improved coordination across the fusion community because too many labs and individuals suffered from the not-invented-here

Fun in Fusion Research. DOI: http://dx.doi.org/10.1016/B978-0-12-407793-5.00015-7

syndrome. Plasma transport, burning plasma physics, and plasma—wall interactions and divertors headed our list. After a meeting in Oak Ridge that included Martin Peng, Ron Parker, Ron Stambaugh, Paul Rutherford, and Harold Weitzner, I wrote a letter to the OFE recommending that they set up a transport task force to bring experimentalists, theoreticians, and computing experts together to improve the understanding of how plasma energy and particles escape the magnetic bottle. I suggested that Jim Callen at the University of Wisconsin—Madison would be a really good person to coordinate this activity. Shortly afterward, OFE set up the activity as proposed. It still exists today (the coordinating role has passed on a few times), and it has a sterling record of improving communications and understanding.

By the end of 1994, I was becoming disenchanted with managing the fusion division. Far too much of my time—more than 75%, even including the many hours I put in outside the workweek—was spent dealing with administrative crap, leaving me very little time to be involved in science. I asked my management if they could find me a job that would be mainly involving science. The lab appointed me to be program director for energy programs. This was a wonderful opportunity, since as far as I know, the ORNL had the broadest energy R&D program of any laboratory in the world—energy efficiency and renewable energies, fossil energy, and fission energy.

I left the fusion program proud of the accomplishments of our program. We had superb collaboration between our experimentalists, engineers, theoreticians, and computer staff, which led to numerous pioneering achievements: cryogenic hydrogen pellet injection; electron cyclotron-assisted start-up; fundamental tokamak, EBT, and stellarator transport studies; energetic particle and MHD theory; diagnostics development; neutral beam development; superconductivity research, including the first test of a high-temperature superconducting coil; materials development; robotics; and plasma systems modeling.

Out of our work on tokamaks and stellartors came Martin Peng's spherical torus (ST), as well as a variety of innovative machines from the stellarator design group discussed previously. Regrettably, we were unable to get approval to build an ST in ORNL, but we sent Martin on a mission to get other labs to build them. He was very successful, and STs sprang into being at Culham and laboratories in Russia, Japan, and U.S. universities, as well as the PPPL. Similarly, in the stellarator area, experiments drawing upon our studies were carried out in Australia, Japan, and Spain.

One other area deserves mention—spin-offs. We were led into areas outside fusion by Hal Hazelton, head of the plasma technology section. Hal had a remarkable ability to spot opportunities and generate new programs. In addition, during the 1980s, when fusion budgets fell, we held the so-called Gong Shows, in which anybody who was involved in our programs could submit a proposal for new work. In turn, we offered a small level of support to help the winners market their ideas.

We received about 60 proposals in the first show, of which around 15 were chosen. We didn't choose to support finding out whether bees communicated using radio frequencies and a few other off-the-wall suggestions. However, we did support using cryogenic CO_2 pellets for cleaning surfaces, improved neutral beam plasma sources for semiconductor etching, and microwave and radio frequency waves for a variety of applications. After a year, our annual spin-off budget was in the range of $5−7 million. Funding came from using pellets to clean paint off F-15 fighters; from Sematech for improved plasma sources to etch semiconductors; and for using various waves to pretreat wood so that less chemicals and waste would be involved in paper manufacturing.

In the general energy area, mine was a gadfly role that allowed me to learn about all the other areas of energy R&D, while spending about 20% of my time involved in fusion. It was sobering for me to realize how ignorant I had been about the other areas. I had adopted many of the prejudices of my fusion colleagues: namely, the scope for energy efficiency improvements is very limited; renewable energies other than hydropower are a joke; fossil can never be cleaned up; fission will be severely limited and is way more polluting; and so on. The reality is that all energy sources pollute, and the differences when everything is taken into account are much more modest than most people realize. Of course, long-lived nuclear waste is an issue, but in a country like France, where more than 70% of the electricity is nuclear, properly separated, and embedded in glass, the high-level waste would amount to only one softball-sized sphere per person during his or her lifetime.

In 1996, I was also appointed to be the executive director of the Joint Institute for Energy and Environment at the University of Tennessee—supported by ORNL, Tennessee Valley Authority (TVA), and the university. This allowed me to get involved in exciting areas such as the issues of air pollution in east Tennessee and the Smoky Mountains and the mind-boggling problem of farm animal manure. You probably don't know this, but the total wet weight of manure produced annually in the United States is around 1.4 billion tons—4 tons for each American.

During this period, I also became chairman of the DOE's Fusion Energy Sciences Advisory Committee (FESAC), a role I held until 2000. This brings me to the pleasure that I got from serving on committees with that wise and witty man, Marshall Rosenbluth.

At a meeting of another committee, we listened patiently as our chair spent an inordinate amount of time clarifying what each speaker had just said. After a particularly egregious clarification, following a coherent and totally clear talk, Marshall raised his hand.

"You have a question, Marshall?" the chairman asked.

"No, but I think it's necessary to explain to the committee what you were trying to say." Marshall then repeated exactly what the chairman had said. I and the other committee members found it hard not to laugh.

"Thank you for the clarification, Marshall," the chairman said with a blank look.

FESAC, like all such government advisory committees, meets in public and has to allow comments from anybody who wants to make them. On one occasion, the committee secretary presented me with a list of about 15 people who wanted to speak. I'd allowed an hour or so for all of them to give us their views and told them that they would each get 5 minutes, including questions. I also said that, if they had viewgraphs, they could only have four.

We were doing fine until the eighth speaker. He had a stack of viewgraphs and after 4 minutes, I had to say something.

"You'll need to wrap up right now or there'll be no time for questions."

"But I'm not even halfway through." He held up a sheaf of viewgraphs.

"You heard the deal: four viewgraphs, 4 minutes, and a minute for questions."

"But what I want to say is important!"

"So are everybody else's comments. We're wasting time. You'll have to stop."

"It's not fair."

"You knew the arrangements." I turned to the committee. "Anybody got questions?"

Marshall raised his hand. "I do."

"Go ahead."

He pointed to the speaker. "Will you show me what you have in the rest of your viewgraphs?"

"Marshall ..." I shook my head resignedly. I knew when it wasn't worth arguing.

Marshall smiled at me urbanely and then asked the speaker a lot of questions.

International Fusion

The International Thermonuclear Reactor (ITER) started under another name, INTOR, during the 1970s and became ITER—Latin for "the way"—in the 1990s. I was on the international committee to define ITER's objectives with others from the United States, European Union, Japan, and the Soviet Union (soon to be Russia). Our chair was Dimitri Ryutov, who was director of the fusion laboratory in Novosibirsk. We held our first meeting at the September 1992 fusion conference of the IAEA in Würzburg, Germany, and argued whether or not we should claim that the ITER D−T tokamak would ignite (i.e., $Q = \infty$). The U.S. and Japanese view was that steady-state operation with $Q \sim 10$ was more important, but the Europeans and Russians were adamant that ignition had to be the main goal.

Interestingly, the Russians asked my ORNL colleague, Jeff Harris, to interpret for them. Besides being a gifted physicist, Jeff was a master at languages and spoke fluent Russian, Japanese, and French. On our way to Würzburg, as we passed through Frankfurt Airport, a young Russian woman stopped and asked in halting English where she could find Aeroflot. She staggered back when Jeff replied in perfect Russian, "Up the stairs over there, and to your right."

Jeff acquitted himself well at the meeting, not simply translating the Russian, but explaining the subtleties of their choice of words.

Because we couldn't agree on all of the objectives for ITER, the ITER Council, chaired by Mikhail Gorbachev's science advisor, Evgenii Velikhov, asked us to try again. In October, we met in Milan.

After 2 days, we had not reached agreement, and in frustration, Dimitri said, "Gentlemen, think glasnost and perestroika. In December, I have to go before the ITER Council in Moscow and give our recommendations. If you can't agree, I will organize another meeting in Moscow ... in December. Be prepared to bring warm clothes."

We compromised on the dual objectives of ignition and steady-state operation at $Q = 5$.

Within the year, an ITER design organization was set up with design sites in Garching, Germany, and Naka, Japan. In an unwise move, they located the supervising site thousands of miles away in San Diego—isn't the politics of international ventures wonderful?

Paul Rebut left JET to be director of the ITER design phase. I was appointed as one of the U.S. experts to the Technical Advisory Committee, along with Bob Conn from UCLA (later UCSD) and Paul Rutherford from Princeton. Various other U.S. experts were involved as the years passed (Figure 15.1).

We met for 3 days at a time, typically three or four times a year (rotating between the United States, Europe, Japan, and Russia), and answered technical questions from Velikhov's committee. We worked in English, and it was necessary to have detailed agreement on the wording of our response before we left. We never failed in many years of meetings, a result of the superb chairmanship of Paul Rutherford.

Figure 15.1 Front row (left to right): the author, Ken Tomabechi (Japan), Paul Rutherford (United States), Boris Kadomstev (Russia), Paul Rebut (ITER director), Robert Aymar (Europe).

Figure 15.2 (Left to right) The author, Paul Rebut, Tihiro Ohkawa, and Dimitri Ryutov.

The toughest meeting was in St. Petersburg during one summer, the time of White Nights, when it never gets dark. During the day, we were in a darkened room with heavy black velvet curtains used to exclude the light so that we could see the viewgraphs projected by a weak lamp. At night, the thin curtains of our hotel rooms let in the outside sunlight, keeping me awake. It was tough to keep our eyes open during the meetings, even for the Europeans, who were nearly in the same time zone; for those of us from Japan and the States, it was impossible. I don't remember how we got out a report. Nevertheless, the Russians were thoughtful hosts, and I do recollect that we had a wonderful picnic in a park one evening.

In 1997, the Engineering Design Activity was coming to an end, and both Technical Advisory Committee (TAC) and FESAC received the final design report to review. While the reviews were generally favorable, there was a growing concern among the partners that the machine was too ambitious and too expensive. It is hard to remember all of the meetings that I attended, but at one of them in San Diego, someone took the photograph shown in Figure 15.2—the occasional glass of wine helped make the activities run more smoothly.

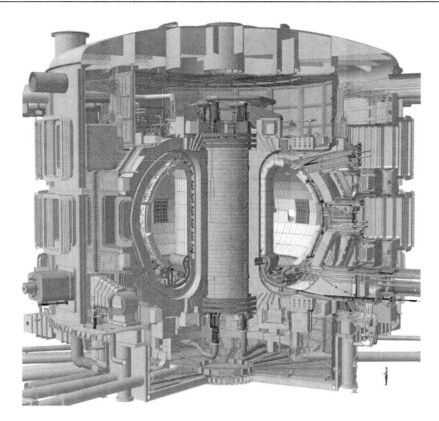

Figure 15.3 The ITER tokamak.
Source: Courtesy of the ITER organization.

In 1998, an agreement was reached among the partners to study smaller, lower cost options with reduced performance requirements. By now, Bob Aymar was the director, and we all tacitly understood that a goal was to halve the construction budget. I was a member of an international Special Working Group, set up to help quantify the new objectives. In the end, everyone agreed that, as had been the case for CIT, ignition ($Q = \infty$) was too ambitious. We settled for $Q = 10$, which allowed the plasma current to be decreased from over 20 MA to around 13 MA (Figure 15.3).

16 What About Fusion Energy?

I was at a meeting where government researchers were reviewing projections for energy demand. Great talks, until we got to this speaker who put up a viewgraph that included a table that was obviously full of errors. A number of us tried unsuccessfully to point out the problems, but the speaker ignored us.

"My numbers are correct," he snapped. "You just don't like what they say." He only glanced at his viewgraphs, confident that they were correct.

Most of us had given up when a colleague from the Argonne National Laboratory, wiser than the rest of us, managed to get his attention by saying in a slow and deliberate manner, "Mister Speaker, would you be prepared to consider the possibility that there are errors in your viewgraph?"

Silence. The speaker looked as if he might launch into another tirade; however, this time, he hesitated and actually studied the viewgraph. More silence as he went over it again. "My God, this is all wrong. The numbers don't add up. Where did this viewgraph come from?' "

He made some compensatory remarks about what he had meant to convey and segued into his next viewgraph. He never apologized. I suppose he felt his admission of the error was enough.

Sir Walter Raleigh Selling Fusion Energy

At a meeting of the fusion community in Snowmass, Colorado, I was asked to comment on selling fusion research to governments and the public.

Now, Bob Newhart is one of my favorite comedians, and I particularly love his telephone monologue in which Sir Walter Raleigh calls the court of Queen Elizabeth I to tell a functionary about tobacco. Bob Newhart played the role of the functionary, answering the phone call. With apologies to Bob, here is what I said, presented as a script:

FUNCTIONARY: (*Pretending to hold a phone to his ear*) Court of Queen Elizabeth, who's speaking? (*He listens and puts his hand over the phone.*) It's nutty Wal. (*He removes his hand.*) What have you got for us this time, Wal? We tried sticking those burning leaves in our ears. It didn't work for us.

Pause.

FUNCTIONARY: Uh, uh. I see. It's a boundless energy source ... fusion, you say. What's fusion?

Fun in Fusion Research. DOI: http://dx.doi.org/10.1016/B978-0-12-407793-5.00016-9

Pause.

Like the sun, but the fuel comes from water ... we've got a lot of that ... Uh, uh ... How hot did you say? ...

He makes a face at the audience. Hotter than the sun. That's pretty hot, Wal. Won't it set the curtains on fire? ...

FUNCTIONARY: (*After listening*) The hot stuff's in a metal vacuum chamber ... think space. I'm trying, Wal... It won't hit the chamber because of the magnetic field ... you're going to surround it with lodestones? ...

FUNCTIONARY: (*Raising an eyebrow*) The magnetic fields will be made with coils ... at what temperature? ... Absolute zero ... (*Raising the other eyebrow*) I know, think space again.

FUNCTIONARY: (*Putting his hand over the phone*) Friends, I think Wal's finally lost it.

(*Speaking into the phone*) Let me get this straight. In the middle, you have this hot ... whatever, like the sun ... and outside it you have something as cold as Iceland. It doesn't sound like a smart thing to do.

Pause.

FUNCTIONARY: (*Looking resigned*) There's more? ... Between the vacuum vessel and these coils, you'll have metal boxes containing lithium at hundreds of degrees. The heat will be used to make steam and power electrical generators. This lithium, I think I've heard of it ... Bipolar people take it to calm down. I hear you, Wal. You've tried it ... Better than the burning leaves. I believe you.

FUNCTIONARY: (*After listening for a while*) So you want me to get the Queen's support. How much are we talking about? ... 150 million gold doubloons. Do you have any idea how many ships that would be? ... Look, Wal, I'm sorry, but ... You have another way of doing fusion? ...

FUNCTIONARY: (*Raising an arm in supplication*) "Okay, I'll listen ... Imagine a small cylinder with a hole at each end ... Size of a thimble... Inside, there's a tiny sphere containing the fuel ... It's near absolute zero. What about the temperature of the sun bit? ... You're going to zap it with powerful beams ... far more than all the guns of the Armada put together ... There's more ... I knew it! You're telling me that round this contraption you've got lithium ... flowing this time? And you're going to zap a few of these things a second. How about the price? ... Same ballpark? ...

FUNCTIONARY: (*Turning to the court and making a face*) Never mind, Wal. Don't call us, we'll call you.

When phrased like this, fusion energy can be made to appear absurd, despite the fact that incredible advances have been made in the research. Let's face it— conquering fusion energy is one of the greatest challenges faced by scientists and engineers. One aggravation to the competent researchers has been a succession of outrageous statements by people who claim they have an easy solution for releasing fusion energy—Argentinian fusion, cold fusion, bubble fusion, and whatever's next.

In the 1950s, an Argentinian scientist conned funding out of the Peron government with the claim that he had produced useful fusion energy. A subsequent investigation showed that he was wrong. Nevertheless, it apparently led Lyman Spitzer to consider how he would make fusion work and from that came his invention of the stellarator.

In the main approach to cold fusion, a voltage is applied to a palladium rod that is immersed in deuterated water, i.e., water in which the only hydrogen is deuterium. After a while, deuterium accumulates in the rod, and finally energy is released into the water. A key question is: After many hours, does more energy come out than went in? The measurement involves integrating very small power levels over all this time. Interestingly, this system with ordinary hydrogen was the basis for a 1920s German cigarette lighter. When the rod was exposed to air, hydrogen reacted with oxygen, making the rod hot.

Some researchers claimed that they obtained a measurable increase in energy, which they attributed to deuterium fusion, resulting from subtle effects in the palladium rod when the deuterium pressure became high. In addition, researchers looked for signs of tritium production and neutrons.

Cold fusion does occur, but with an infinitesimal probability compared to hot fusion. It remains an interesting area to investigate, even if it won't lead to useful net energy production. It remains unclear to me why a number of competent scientists were convinced that it was a potential energy source. I could chalk it up to natural enthusiasm and the tendency to dwell on positives rather than negatives, the vision of a Nobel Prize dangling in front of their eyes; or to the possibility of making lots of money.

My colleague, Mike Saltmarsh, at the request of the ORNL, had the unenviable job of looking into both cold fusion and bubble fusion. On one occasion, he asked a researcher the following question: "I understand that you see cases where more energy comes out of the cell than goes into it. Do you see cases where you measure less energy?"

The answer was "Yes, but that's physically impossible, so we ignore those cases."

Come on! How about saying that on average there's no effect, and the differences reflect experimental error? I have also heard that some researchers admit they have positive and negative gains, but the positives do outweigh the negatives. Nevertheless, to me, the claims for gain appear to be based on questionable evidence.

On another occasion at ORNL, the experiment was mounted on a tray in the basement of a building. After the experiment had been running for many hours overnight, the researchers came to check on the experiment. They were thrilled to see that the chart recorder monitoring their radiation counter had risen during the night and was sitting at a significantly higher level. With great excitement, they told a senior scientist what had happened. He went to look.

After inspecting the setup, he said, "I'd like to take the tray upstairs, with the radiation counter connected. Would someone help me bring everything?"

When they got upstairs, the chart recorder dropped back to its resting position. It was then that they realized that the effect had been due to radon leaking into the basement when the air-conditioner was turned off at night!

In bubble fusion, acoustic cavitation is used to create bubbles in a liquid, such as deuterated acetone, which then collapses, raising the temperature. Its history sounds very similar to cold fusion, in which a key lesson was that it is essential to involve a large number of the very best scientists if you are to avoid making errors when you are measuring very subtle effects. As one brilliant scientist said on hearing about cold fusion, "I might believe in one miracle, but this seems to require two."

Electrostatic confinement is one other area of research. In my opinion, sadly, it is unlikely to make useful net fusion energy. It was proposed in the 1950s by Philo Farnsworth, inventor of the basic technology for television. He tested the idea of focusing charged particle deuteron and triton beams in a spherical arrangement. This approach uses electric fields to trap the charged particles. A common arrangement has concentric wire spheres at different electric potentials. Energetic ions may also be fired in radially by ion guns.

This area was studied further by Robert Hirsch, originally at the University of Illinois, who demonstrated that it was possible to make billions of 14 MeV neutrons in a pulse. This approach fell out of favor for fusion energy, as theoretical studies indicated that it would not be a net energy producer because of the scattering of beam particles out of the focused region. The other approaches showed more promise. However, different applications, such as explosives detection using the energetic neutrons produced by fusion, led to a renewed interest in Japan at Kyoto University, and in the United States at the University of Illinois and the University of Wisconsin.

Some years later, Harold Furth of the PPPL was talking to Bob Hirsch, who, at the time, was head of the DOE's Office of Fusion Energy and the person who funded Harold's laboratory.

Bob said, "I know you guys believe magnetic fusion is the answer. But I think it's too complicated and you need to look for a simple system, electrostatic containment for example." He sat back.

"Interesting thought," said Harold, stroking his beard. "You know, we should come up with a catchy name for your approach." He continued to stroke his beard for a short time. "I think I have it. We describe a magnetic containment system as a 'magnetic bottle.' In the same vein, let's call your approach an 'electrostatic crock.'"

I don't want to give the impression that irrational optimism can be found only in the fringes of the main fusion program. One early experiment contained a toroidal vacuum vessel split into four segments. Insulating breaks between the segments allowed a voltage to be applied across each insulated gap (similar to the construction of the Spider shock-wave experiment mentioned earlier). This voltage was used to drive a current and create and heat the plasma. During start-up of this device, the only plasma diagnostic available was a spectrometer to measure the visible light emitted by the plasma. After the team had operated the experiment for a number of pulses, they brought their leader (i.e., the inventor of the concept) the Polaroid photographs from the spectrometer.

"There's nothing on the photographs," one of his staff said cautiously. "Maybe it isn't working."

"Let me think about that," the leader said. When he called them back, he stated, "The explanation is simple. The plasma is so hot that it is not emitting visible light. Go on firing shots while I work out how to make a measurement."

The staff dutifully continued operating the experiment, piling up Polaroid pictures of nothing. Finally, they went to their boss and told him, respectfully, that they did not think there was any current inside the vacuum vessel, and therefore, no plasma.

"You have no faith," he replied. "I will show you. Split the vacuum vessel and put a sheet of Kleenex across the gap. Then seal it up again. After you fire, you will see light from the burned paper."

After they had performed the test, the bravest staff member imparted the news, "There's still nothing on the Polaroids."

"It must be hotter than I thought," the leader replied. "Open it up. You'll find a large hole burned through the paper."

No hole! But the staff did find something else: there was inadequate insulation between the metal segments. The current had flowed only in the metal of the vacuum vessel. When this problem was fixed, the experiment performed well and led to an enormously successful program.

The fusion program that I have experienced has been full of similar surprises and ups and downs. Nevertheless, there has been a steady progress. I have the knowledge, as I engage in my second 50 years in the field, to state with confidence that eventually, if we continue a collaborative, worldwide effort, fusion energy will be exploited for the benefit of the world.

17 Fusion and the Universe

Forces in the Universe

What role has fusion played in the universe? After the Big Bang, gravity dominated the behavior of the universe during the first 10^{-32} seconds, when the temperature was a huge 10^{23} eV.[1]

Subsequently, the electro-weak and strong nuclear forces became important. The weak interaction split into two parts—the weak interaction and the electromagnetic force—around 10^{-11} seconds when the temperature had dropped to 10^{12} eV.

After less than a second, when the temperature reached 10^{9} eV, the strong force produced protons from the fusion of quarks—the tinier building blocks of other particles. At about 1 second after the Big Bang and a temperature of 10^{7} (10 million) eV, the forces operating in the sun and other stars fused protons to produce the light elements deuterium, tritium, and helium.

Fusion also produced and still produces the heavier elements—up to iron—in the dense material of stars, where the temperature is typically less than 10^{6} (1 million) eV. The even heavier elements, including uranium, are produced during supernova explosions.

At less than about 1 eV (11,600 K), atomic forces take over during the interactions of electrons and charged ions (the building blocks of atoms and molecules). They are involved in ordinary chemical reactions.

The point of this discussion is to show that the origin of all the energy resources available to us on Earth comes from nuclear fusion in the core of our sun and other stars.

The light elements deuterium and lithium are the closest to being exploited to produce fusion energy for peaceful purposes. The evidence for the formidable amounts of energy in such fuels was made clear in hydrogen bomb tests. The heaviest isotope of hydrogen (tritium) is radioactive and decays, but it may be produced by bombarding lithium with neutrons. Deuterium–tritium (D–T) fusion occurs at about 100 million degrees (10^{4} eV). Deuterium, the second isotope of hydrogen, exists as one part in 6500 in all hydrogen. It is, in effect, limitless in the oceans, but reasonably priced lithium is not a limitless resource.

---

[1] $10 \times 10 \times 10 \times 10 \times 10 \times 10$ (i.e., 6 times) is denoted by 10^{6}. I millionth $1/10^{6}$ is denoted by 10^{-6}. Consequently, $10^{-32} = 1/$a hundred million million million million million million. $10^{23} =$ a hundred thousand million million million.

Fun in Fusion Research. DOI: http://dx.doi.org/10.1016/B978-0-12-407793-5.00017-0

Table 17.1 The Highest Cross Section Fusion Reactions

Fusion Reaction (1 keV = 11.6 million °C Absolute)	Energy Produced (MeV)	Average Ion Temperature Required (keV)
D−D → T(1.01 MeV) + p(3.02 MeV)	4.03	30−50 keV, including fusion product reactions
50% → He^3(0.82 MeV) + n(2.45 MeV)	3.27	
50% D−T → He^4(3.5 MeV) + n(14.1 MeV)	17.6	10−20 keV
D−He^3 → He^4(3.6 MeV) + p(14.7 MeV)	18.3	30−50 keV

D, Deuterium; T, tritium; He^3, helium-3; He^4, helium-4; n, neutron; p, proton.

Deuterium−deuterium (D−D) fusion occurs but requires a temperature of around 400 million °C (see Table 17.1). The fusion of deuterium produces helium-3 and tritium, which may be recycled to make the deuterium fusion more effective. A more speculative option is to mine helium-3 on the moon and use it in the deuterium−helium-3 cycle. In my opinion, a better solution is to produce tritium by D−D fusion, extract it (a tricky task), and let it decay to helium-3. Some consideration has been given to more exotic fusion fuels, such as protons and boron, but such fuels require even higher temperatures, and I suspect that it will not be possible to extract useful energy from them.

As for cold fusion and bubble fusion, frankly, they have been oversold as energy sources.

Thus, in the very long term, energy for the Earth will be provided one way or another by fusion—the sun and deuterium.

Unfortunately, owing to rash statements over the years about how close we are to harnessing fusion energy (often from people who were not the best informed), it has been described as a dream that will always be 25 years away. In reality, progress has been steady over the last few decades, and given the actual budget expenditure, the program has met many of the goals stated earlier; including reaching temperatures of 600 million °C in laboratory experiments.

Fusion Power Plants

Figure 17.1 shows the principal features of a D−T power plant.

While the figure shows a magnetic fusion plant, an inertial fusion plant would have a similar setup without the coils but with drivers. In each case, the fuel is heated and confined (magnetically in MFE, and inertially in IFE) in a vacuum chamber. The heating power and fusion power end up on the first wall. The neutrons pass through the wall and are collected in a blanket that contains lithium.

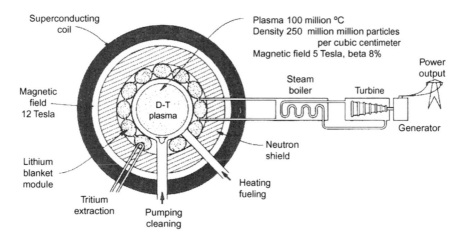

Figure 17.1 Diagram showing the components of a magnetic fusion power plant.
Source: Courtesy of ORNL.

Tritium, which does not occur naturally, is produced by neutron—lithium reactions and continues the fueling along with additional deuterium. The captured heat is converted to electricity (P_E), and a fraction η_R is recirculated to produce, heat, and confine the fuel and to operate the conventional balance of plant—cooling, instrumentation, etc.

An important issue is the radioactivity induced in the walls and blankets by the neutrons. Nevertheless, unlike a fission plant, a fusion plant does not have extremely long-lived isotopes. By careful choice of materials, the induced activity would decay on a timescale of 100 years to a level hundreds of thousands of times less than for fission, with no need for storage of waste over the geological time periods contemplated for repositories, such as Yucca Mountain in Nevada. Many experts believe that a form of ferritic steel, improved to handle the neutrons without outrageous swelling, will be the main structural material. In addition, silicon carbide composites may play a role in the blankets (Figure 17.2).

Fusion Energy and the Plasma State

In recent years, great progress has been made in creating and understanding fusion plasmas and in the generation of fusion power in the laboratory. As stated in the 1999 document prepared for DOE by the Fusion Energy Sciences Advisory Committee, *Opportunities in the Fusion Energy Sciences Program* (http://www .ofes.doe.gov/more_html/FESAC/FES_all.pdf).

Today, there is little doubt that fusion energy production is feasible. The challenge is to make fusion energy practical. As a result of the advances of the last few years, there are now exciting opportunities to optimize fusion systems so that an

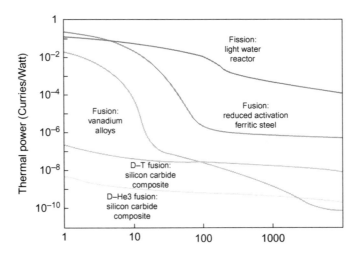

Figure 17.2 Radioactive decay in the materials used in fusion and fission power plants. *Source*: Courtesy of the U.S. DOE-OFES.

attractive new energy source will be available when it may be needed in the middle of the next century (twenty-first). Among the reasons to pursue fusion energy research are: the risk of severe environmental impacts from some existing methods of energy production; and the risk of conflicts arising from energy shortages and supply cut-offs. Fusion is a scientific and technological grand challenge. It has required the development of the entire field of high-temperature plasma physics, a field that contributes to the description of 99% of the visible universe.

The fusion reactions that occur at the lowest temperature with the highest probabilities, and might lead to a commercial power plant, are listed in Table 17.1.

The majority of work in fusion energy research concentrates on the D−T cycle because it has the highest reaction rate and lowest temperature requirement. However, ultimately it is the D−D cycle that could produce unlimited energy on earth. Our sun will expand to engulf the solar system before we could use up all the deuterium in the oceans. In addition, the 14 MeV neutrons produced by D−T present a huge problem for the materials that surround the plasma (walls) because of the damage they cause (more about this later in this chapter). The lower energy neutrons from D−D are less of a problem.

D−He^3 is an interesting additional option because on its own, it produces no neutrons. Using a low percentage of deuterium would reduce the neutrons from the D−D reaction and significantly reduce the wall damage. Helium-3 does not occur in large quantities on earth.

An alternative option would be to run a D−D system and extract as much of the tritium as possible before it reacts. Tritium decays to helium-3, with a half-life of about 12 years. If this helium-3 were recycled, it would supplement the helium-3 produced by the D−D reactions and lead to a quasi D−He^3 system. Calculations indicate that this system might have up to seven times less neutron damage than an equivalent D–T system.

Thus, in fusion energy research, the goal is to contain a high-temperature mix of light elements such as deuterium, tritium, and helium-3, and their charged fusion products long enough that the energy released in exothermic fusion reactions exceeds the energy used to contain and heat the mixture.

At temperatures above 10,000°C, all materials are in a state known as a *plasma*— a state in which electrons are stripped from atoms and the resulting material has equal numbers of free electrons and positive ions—the sun and most of the visible universe, flames, arcs, fluorescent lights, and the glow discharge used in etching semiconductor chips.

Plasma is sometimes called the fourth state of matter—solid→liquid→gas→ plasma; the changes occur as the temperature is raised, as discussed in Chapter 1.

Two main approaches have been pursued vigorously for containing this high-temperature state of matter, and in turn, each has two major subareas:

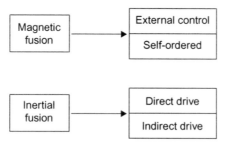

Background to MFE

MFE uses magnetic fields, which have the property that the electrons and ions spiral around the field lines and, in principle, may be isolated from material walls. The early efforts to produce fusion plasmas used cylindrical systems. The first such attempt was made in the 1930s by Arthur Kantrowitz in a laboratory at NACA, the forerunner of the National Aeronautics and Space Administration (NASA). He submitted a patent application in 1941.

It was soon realized that such systems suffered badly from end losses, even when very high magnetic fields at the ends constricted the loss region—called a *magnetic mirror*. Nevertheless, the early research on systems that didn't pan out provided valuable insight into plasma behavior.

The main approach used to eliminate the problem with the ends of linear systems was to bend the configuration into a circle inside a toroidal vacuum vessel, as shown in Figure 17.3.

The total magnetic field spirals around the torus, producing magnetic flux surfaces, nested like the layers in an onion. The particles diffuse across surfaces owing

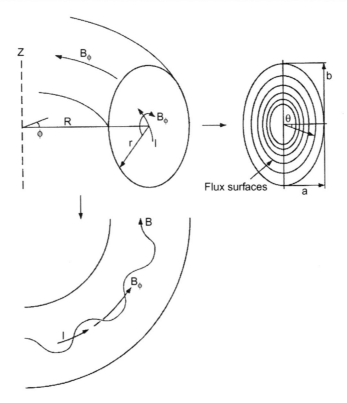

Figure 17.3 Nested flux surfaces in a toroidal system, produced by toroidal ($B_\circ$) and poloidal (B_θ) magnetic fields. The charged particles tend to stay near these flux surfaces, giving good confinement.
Source: Courtesy of ORNL.

to collisions and the effect of instabilities that provide fluctuating fields and density.

In the earliest such experiments, the magnetic field in the poloidal direction (B_P) was provided by driving a toroidal current that also produced and heated the plasma—the *diffuse pinch*. In the late 1940s, Sir George Thomson patented a diffuse pinch fusion reactor in Britain. His 1951 patent contained many of the features found in today's design studies of fusion power plants—e.g., a 4 m major radius and the current drive (Figure 17.4).

Unfortunately, systems such as the diffuse pinch are unstable, and in the 1950s, three solutions were proposed to overcome the problem:

- In Russia, Igor Tamm and Andrei Sakharov added a large toroidal field ($B_T > B_p$), the *tokamak*.
- Roy Bickerton at Harwell in the United Kingdom suggested adding a weaker toroidal field ($B_T > B_p$), leading to the *reversed field pinch*, so-called because the magnetic field

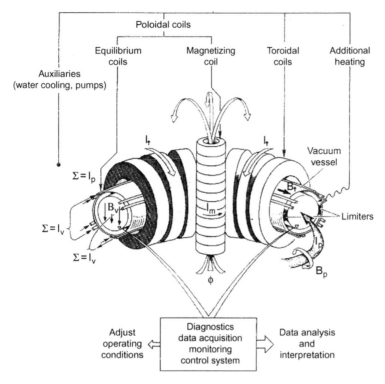

Figure 17.4 A cutaway drawing of a tokamak illustrating the principal features.
Source: Courtesy of ORNL.

spontaneously reverses, leading to a system with improved confinement. This approach was also pioneered at the Los Alamos National Laboratory.
- *The third proposal came from Lyman Spitzer at Princeton, who proposed using twisted external coils to produce such a twisted field, the *stellarator*.

In all of these systems, the magnetic field spirals around the toroidal plasma producing nested magnetic flux surfaces; such systems can have good plasma properties. Later still, proposals were made for more compact magnetic fusion configurations, based on earlier discoveries, the *field reversed configuration* and *spheromak*.

In the early days, the focus was almost totally on achieving good confinement with fusion-relevant temperatures, which back then meant achieving around a kilovolt.

In fact, it would not be until the end of the 1960s that fusion would show real promise, when the Russian tokamak, T-3, achieved an electron temperature of 12 million °C.

In 1971, buoyed by the success of the world's tokamak programs, the European Union's fusion program initiated a study of options for a major next step tokamak— the JET. In 1972, the Luc committee, with members from the laboratories partnering in the European effort, recommended that a tokamak with 3 MA of plasma current should be designed. At this time, the world's largest operating tokamaks had far less than 1 MA. The reason for the choice was simple: At this current, the majority of the alpha particles produced by the fusion of deuterium (D) and tritium (T) would be contained in the plasma and slow down by collisions heating the plasma—the goal of a self-sustaining fusion system. So, the proposed device would have not only a huge current but also operation with D−T—bold steps.

Shortly afterward, Japan (JT-60), the Soviet Union (T-20, in the end not constructed), and the United States (TFTR) took a similar step. Ultimately, ion temperatures of 500−600 million °C were obtained in these tokamaks.

The encouraging advances through the 1990s in the tokamak area led, eventually, to approval of construction of the ITER in the south of France. ITER is designed to use D−T plasmas and produce up to 400 MW of fusion power for 500 seconds, with lower fusion power in steady state. This activity is a joint effort of the European Union, India, Japan, South Korea, Russia, and the United States.

Steady progress has been made in alternative configurations:

- The stellarator was demonstrated to have a performance comparable to that of the tokamak through research programs in Germany, Japan, and Russia, but without the problems associated with having a plasma current. Complementary work in Europe and the United States improved the understanding of the area and expanded the opportunities for an attractive power plant. The primary issue is to optimize the coil configuration for plasma performance and simplicity. The superconducting coil, LHD in Japan, has a performance comparable with intermediate-size to large tokamaks.
- A similar-scale superconducting stellarator with modular (twisty toroidal) coils WVII_X is under construction in Greifswald, Germany.
- Work on the reversed field pinch in the University of Wisconsin−Madison and the University of Padua, Italy, has led to a significant improvement in control of the plasma and of confinement.

Background to IFE

IFE compresses and heats a sphere of fusion fuel, which has time to fuse, owing to inertia, before it expands.

The two main approaches to IFE involve the direct and indirect drive of the target. In the latter case, the incident beams interact with material in the box (called a *hohlraum*) that surrounds the target, producing X-rays. The X-rays, created to bathe the target uniformly, ablate the surface compress and then heat the central hot spot, leading to ignition and burn (Figure 17.5).

The incident power may be tailored in a variety of ways to optimize the compression and heating. For example, in fast ignition, the main beams (nanoseconds) serve only to compress the target. A second, very short pulse beam (much shorter than a nanosecond) then provides the hot spot. In a variant of this approach called *shock ignition*, the beams are shaped to provide rapid shock heating at the end of the pulse.

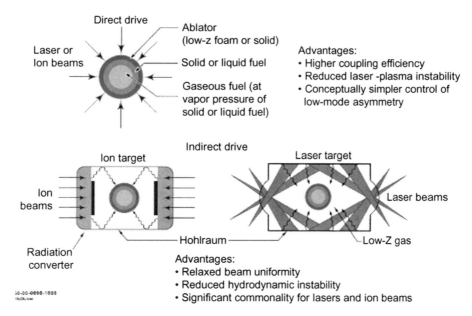

Figure 17.5 Direct and indirect drive options.
Source: Courtesy of LLNL.

Today, the bulk of the fusion energy research is focused on MFE and IFE, with a very small effort put toward alternatives that combine elements of the former approaches (e.g., rapid compression of a magnetized plasma). The dramatic effect of hydrogen bombs is evidence that it is possible to release fusion energy using inertia to hold a compressed and heated capsule together long enough for substantial fusion to occur. The efforts to develop a system for peaceful uses started with the development of powerful lasers in the 1960s, in an effort led by Academician Basov in the Soviet Union and John Nuckolls at LLNL in the United States. The first experiment at LLNL consisted of 12 beams of a ruby laser called 4-pi, aimed at the center of a spherical vacuum vessel. Shiva, the subsequent laser built in 1977, had 24 beams and delivered a 0.5−1 ns pulse of 10.2 kJ of infrared light at a 1062 nm wavelength or smaller peak powers over longer times (3 kJ for 3 ns) (Figure 17.6).

Since that time, larger and larger single-shot (i.e., once an hour or so), low-efficiency (<1%) neodymium glass lasers have been built at LLNL and at laboratories in Russia, Europe, and Japan. The NIF (shown in Figure 17.7), the newest laser at LLNL, has 192 beams and can deliver up to 2 MJ of ultraviolet light energy at 3500 Å.

A similar system, the Laser Mégajoule, is under construction in France. These systems are expected to be able to ignite a capsule and produce energy gain:

(Fusion energy out)/(Laser energy in) > 1.

Figure 17.6 Shiva laser amplifiers and target chamber.
Source: Courtesy of LLNL.

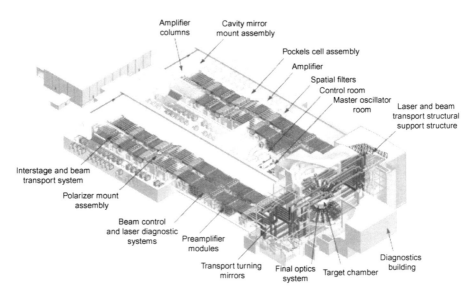

Figure 17.7 NIF.
Source: Courtesy of LLNL.

Recently, progress has been made at LLNL (improved glass laser) and at the NRL (NIKE, a krypton fluoride gas laser) in developing lasers relevant to the fusion goal that have higher efficiencies (5–10%) and can pulse a few times a second. An alternative approach would be to use heavy ion beams to compress and heat the capsules. These accelerator-based systems have the advantage of better efficiency (~30%) and an inherently high repetition rate.

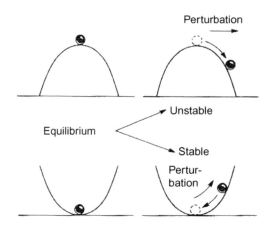

Perturbation

Equilibrium

Unstable

Stable

Pertur-
bation

Figure 17.8 A ball in the bottom of a bowl is stable because if we displace the ball, it will return to the bottom under the force of gravity. The other way around, the ball will fall off.
Source: Courtesy of ORNL.

Main Challenges for Fusion Energy Research

Achieving a Stable Plasma

It is easy to make plasma; simply take any material and pour in energy (e.g., by discharging a large current through it—lightning and arcs). The problem is that without other actions, the plasma will not hang around. It will be unstable. A commonly used analogy of how difficult it is to control plasma describes it as trying to contain loose Jell-O on a tennis racket. Pictures of the turbulent surface of the sun show jets of plasma being expelled into space. At home, the flickering of the plasma in a neon light is a gentler example of instability. In general, whether a situation is stable or unstable may be formulated in terms of the concept of a potential well (Figure 17.8).

A common cause of instability, explained by Rayleigh and Taylor, occurs when we try to put a layer of high-density liquid on top of a lower density one. As shown in Figure 17.9, any small downward perturbation of the interface will grow because the force of gravity on the higher density material will cause it to drop further. Eventually, the liquids will flip so that the lighter one is on top—this effect can be seen in a bottle of vinaigrette.

In the case of the sun, the force of its gravity is sufficient to hold it together, even though particles that gain enough energy from instabilities can escape—the solar wind.

A critical parameter in MFE is beta, which is the ratio of the plasma pressure to the magnetic pressure. From an economical point of view, the pressure needs to be about 1 atmosphere to get a high enough fusion power density in a burning D−T plasma, while the magnetic field should be as low as possible to keep the magnet system cost down. Typically, we need beta to be less than or equal to 5%.

In the so-called magnetic bottle, the question for any configuration is whether the plasma is on the inside or the outside of the bottle. In the case of the simple

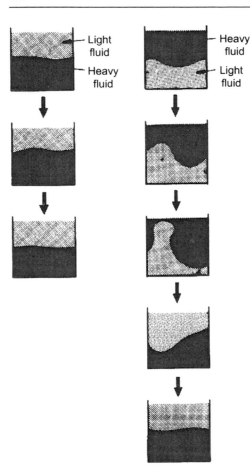

Figure 17.9 Rayleigh–Taylor instability. *Source*: Courtesy of ORNL.

mirror, the magnetic field decreases away from the plasma in the median plane, and the plasma pushes its way out. The mirror ends have a field increasing away from the plasma, but, in total, the system is unstable. The problem may be overcome by adding additional fields but, to date, no one has demonstrated a way to block particles and energy leaking out of the ends.

In the simple pinch, the plasma current thrashes around like an unconstrained fire hose and hurls plasma into the walls. This happens when the current channel kinks slightly, because its magnetic field bunches (becomes stronger) where it bows in and spreads out (becomes weaker) where it bows out, causing the kink to grow (Figure 17.10).

In IFE, any nonuniformity in the spherical surface of the capsule will grow during compression owing to the Rayleigh–Taylor instability. Nonuniformities must be small enough (typically they are held to one part in a million on the radius), so that the symmetry of the compression is not destroyed.

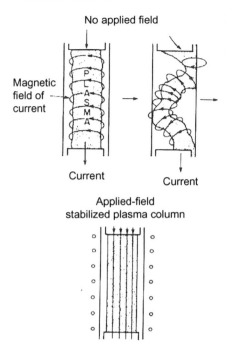

No applied field

Magnetic
field of
current

Current   Current

Applied-field
stabilized plasma column

Figure 17.10 Stabilization of plasma using an
axial magnetic field.
Source: Courtesy of ORNL.

Confining the Plasma Energy

Numerous fusion systems have been developed that avoid gross instability. The
next question is whether they will contain the heat well enough so that the temperature can be sustained (or increased in the case of IFE) by the alpha heating. In magnetic fusion, some auxiliary heating may be used to maintain the plasma
temperature, provided the overall energy gain is high enough. Heat will be lost in
all circumstances because of collisions that lead to particle and energy diffusion
and also because of various kinds of radiation—bremsstrahlung, synchrotron, and
line. Two additional problems increase the heat loss:

1. Instabilities can lead to fluctuations in density, electric field, and magnetic field that
   enhance the diffusion of particles above the classical collisional level.
2. Nonhydrogenic impurities (crap) from the walls can increase the radiation losses.

   Improving energy confinement was the main focus during the first decades of
fusion research. A deep understanding of plasma behavior was achieved through a
combination of theory, experiment, and modeling ... and some serendipity.

The Lawson Criterion

A simple formula that encapsulates the requirements to achieve sustained fusion
was derived by John Lawson at the Rutherford-Appleton Laboratory. For a D−T

system with an ion temperature in the range 10–20 keV, the alpha particle heating in a magnetic confinement plasma compensates for the rate at which heat is lost when the criterion may be written as

$$\text{Density} \times \text{Temperature} \times \text{Confinement Time} > 3 \times 10^{21} \text{ m}^{-3} \text{ keV s}$$

Most magnetic fusion reactor designs aim for a density $\sim 10^{20}$ m^{-3} and a confinement time of a few seconds. For a catalyzed D–D system (one that allows the reaction products to fuse), the criterion is about 10 times higher because of the lower fusion reaction rate.

In contrast to MFE, the ignition condition in IFE is achieved with confinement times around a billionth of a second and densities (pressures) correspondingly higher. The burn of an ignited fuel mass is generally quenched by hydrodynamic expansion. A rarefaction wave moves in from the outside of the fuel at the speed of sound. By the time this rarefaction has moved a fraction of the radius r, the fuel density has dropped so low that it no longer burns efficiently. The fractional burn-up ϕ is

$$\phi = \rho r / [\rho r + 6 \text{ (g/cm}^2)] + N_0 \tau / [N_0 \tau + 5 \times 10^{15} \text{ (s/cm}^2)]$$

where ρ is the matter density in the fuel, τ is the confinement time, and N_0 is the particle number density. The targets rely on central ignition followed by propagation of the burn via alpha deposition and electron conduction into the surrounding cold fuel. Once the fuel reaches 10 keV with a ρr equal to the range of fusion alpha particles (~ 0.3 g/cm^2 at 10 keV), the burn will propagate and ignite cold fuel.

In MFE, subtle improvements in the magnetic configuration and better control of the plasma have led to conditions in tokamaks and stellarators in which the confinement of ion energy is classical and impurities are at an acceptable level.

In IFE, systematic improvements in capsule design and laser power and quality have led to steady improvement in fusion energy production.

Heating and Fueling the Plasma

In the earliest experiments, the plasmas were created and heated by current—known as *ohmic heating*. This approach in a system with good confinement led to an electron temperature of 1 keV in the Russian tokamak T-3 in the late 1960s. This approach has advantages (simplicity) and disadvantages, i.e., as the temperature goes up, the plasma resistance drops and the efficiency of heating goes down. Also, the current drives instabilities.

Alternative heating approaches were developed, primarily through experiments in tokamaks and mirrors. Powerful neutral hydrogen beams, developed at ORNL and LBNL, led the way in the United States, producing an ion temperature of 7.5 keV in the late 1970s in the Princeton PLT tokamak, and later more than 50 keV in TFTR. The neutral hydrogen atoms can cross the magnetic field, but

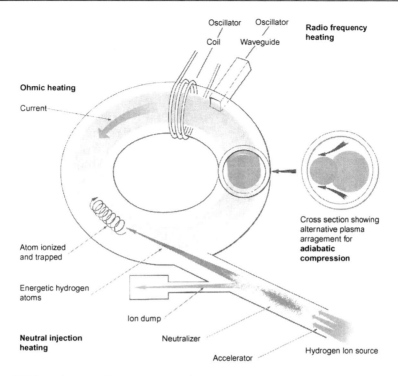

Figure 17.11 Various heating systems used in MFE.
Source: Courtesy of JET.

when they meet the plasma, they are ionized and trapped as charged particles. They then slow down and heat the plasma. The injected atoms must have sufficient energy to penetrate the plasma—ideally, to its center. In practice, this meant tens of kiloelectron volts and hundreds of kilowatts for hydrogen atoms in the modest-scale experiments of the 1960s and 1970s. The energy rose to 160 keV and tens of megawatts in the 1980s, and for ITER, the injection energy is up to 1 MeV (Figure 17.11).

Various generators of electromagnetic waves were also developed during the 1960s and 1970s, ranging from radio waves (megahertz range) to microwaves (gigahertz range). In the past two decades, combinations of these heating systems at the level of tens of megawatts have been used in a variety of magnetic fusion experiments. They have delivered ion temperatures of up to 60 keV and electron temperatures of up to 25 keV.

In parallel, the theory has advanced to the point where computer models can predict accurately the interactions of the heating sources with the plasma.

From the 1930s through the 1960s, fueling of the MFE plasmas was from an initial gas fill, supplemented by puffing in additional gas and by adding particles from neutral beams. A problem with introducing gas at the edge is that it can exchange charges with energetic ions in the edge, allowing them to escape and bombard the walls.

In the mid-1970s, an injector of solid hydrogen pellets was built and tested successfully on the ISX-A tokamak at ORNL. This approach has the advantage that the neutral fuel is deposited well into the plasma.

In IFE, the plasma production and heating to date has been achieved by the action of the laser beams and by X-rays produced when a massive current is discharged through a cylindrical array of fine wires called the *Sandia Z-machine*. In the main approach, the rocket action of the ablation at the capsule surface compresses the target, raising the density to many times the solid level. The trick is to tailor the shape of the laser or beam pulse to delay heating as long as possible, to maximize compression; then, shockwaves converge on the center, heating it to multi-kiloelectron volt temperatures.

An alternative approach developed at LLNL and Osaka University, called *fast ignition*, uses one large laser to compress the capsule and a second, ultrafast laser pulse to heat the center. A variant of this approach pioneered at the University of Rochester—shock ignition—has a spike in power at the end of the compression laser's pulse. Both of these approaches have the possibility of reducing the driver energy for ignition from the megajoule range to less than 1 MJ.

Handling the Fusion Power

Eventually, all the energy put into a plasma to create and heat it comes out and must be handled. In a fusion plasma, there is also the energy created by fusion reactions. The energy leaves in the plasma and energetic ions that escape the confinement, as electromagnetic radiation, and in the neutrons and neutral atoms that escape. The walls of the vacuum containment vessel must be able to handle the heat load (up to 10 or more MW/m^2 on average), erosion by bombarding plasma and energetic particles, and damage caused by the fusion neutrons.

The heat load issues are similar to those encountered in a rocket nozzle, but in MFE, the heat must be handled in steady state and, in IFE, the heat load will be pulsed a few times a second at much higher transient levels. In MFE, magnetic flux is diverted from the plasma edge and the outgoing plasma is directed to collector plates that can cope with the high heat loads—a divertor.

A similar solution has also been studied for IFE, although an alternative option exists in which, in the absence of an applied magnetic field, the walls would consist of a thick layer of flowing liquid lithium. The lithium would stop most of the neutrons and all of the ions, electrons, and radiation from the target explosion. It would also produce tritium while protecting the material wall of the vacuum chamber.

Neutron damage is a material issue that has been solved for fission reactors, but in fusion, the neutrons are much more energetic (up to 14 MeV). They create hydrogen and helium by transmutation, in addition to displacing atoms in the wall by collisions. The hydrogen can leak out, the helium accumulates in the voids created by the displacements, and in most materials of interest (e.g., austenitic stainless steels), unacceptable swelling is caused. Good progress has been made in developing ferritic steels and silicon carbide composites that show promise in being able to deal with a power-plant level of 14 MeV neutron fluence.

To date, the materials research has been done without a powerful source of 14 MeV neutrons. It is planned to build the International Fusion Materials Irradiation Facility (IFMIF) at a new fusion site at Rokkasho in Japan. IFMIF will fire energetic deuterons from an accelerator into a flowing liquid lithium target to produce a spectrum of neutrons around 14 MeV.

Challenges for a Fusion Power Plant

The cost of electricity (COE) of a power plant consists of three parts relating to the following:

- The mortgage on construction—typically, a fixed charge rate (f_r) of $8-10\%$ per year of the capital cost (C_0)
- The annual operating costs (C_a)
- An annual charge to cover eventual dismantling and disposal ~ 0.5 c/kWh

$$\text{COE} = (f_r C_0 + C_a)/(8760 F_a P_e) + 0.5 \ c/\text{kWh}$$

Number of hours in a year $= 8760$, and F_a is the availability (fraction of the year) to operate at the assumed full power level (P_e).

The biggest cost component for an MFE power plant is the magnetic coil system. Fortunately, enormous progress has been made through developments for the world's superconducting fusion devices and accelerators. The state-of-the-art magnets that have been developed for ITER are at the scale and field level (14 Tesla) required for a demonstration power plant (DEMO). In addition, ITER will test heating and diagnostic systems that are, in some cases, prototypical of those required in a DEMO. The weakest area of development is the lack of operating experience in steady state at power-plant heat levels, let alone with the 14 MeV neutrons produced by D−T fusion.

The main cost component for IFE is the driver, and much more development is required to demonstrate a long-lived, multishot-per-second system. In a similar situation to MFE, tests at high power are required of a continuously operated repetitive system.

I believe that all of these challenges will be met in both MFE and IFE: however, availability is the greatest uncertainty facing fusion. Assuming that the physics holds up, and that cost estimates for present power-plant studies are more or less correct, an availability of 70% or more of the year will be required to obtain COE < 10 c/kWh. But today, there is essentially no database for component and system reliability. An advantage for IFE is that the bulk of the driver is outside the radiation zone, allowing hands-on maintenance.

In MFE, this will start to come when ITER operates, but many facilities and tests in parallel will be required to qualify materials and components. A fusion component test facility has been proposed to complement ITER. A similar component test facility has been proposed for the IFE program.

Reflections on 50 Years in Fusion Research

I feel incredibly fortunate to have picked fusion energy as my area of research. While fusion continues to have occasional periods when people are not working well together, generally it has been undertaken in a spirit of collaboration, which is a model for science. This has been particularly the case in the experimental area, where I started my career. Most fusion experiments, going back to the time of ZETA, have involved large numbers of scientists contributing to a common goal. This is because, to function effectively, a fusion device must have large numbers of diagnostics, data acquisition, and heating and fueling systems. An experiment may involve anywhere from 5 to more than 100 scientists.

Of course, there are prima donnas and rivalries. And when they are not contained, they cause damage to the program. When I returned to the United States in 1977, having been working on JET, I was struck by the contentious nature of the U.S. fusion program. At the time, JET involved scientists from 13 countries. Most of us remembered being bombed during World War II, yet we were able to work together very well. At the 1972 EPS fusion meeting in Moscow, one of the German scientists, in order to obtain a visa, had to answer the question, "Have you previously visited the Soviet Union?" He answered no. Yet he had, as commander of a tank at Stalingrad.

One reason that it was possible to set up a European project like JET was that each European laboratory received the bulk of its funding from its own government, and JET was funded by the union as a whole. In contrast, in the United States, each laboratory was and is in competition for DOE funding. For this reason, even though the collaborative JET program had already started, the U.S. TFTR was predominantly undertaken by only one laboratory—PPPL. Their scientists did a great job exploiting the facility, but I have often wondered how much better the experiment would have been had it been a national effort.

It took many years before the need for getting all the players involved became the norm in our programs. Nowadays, nearly all of the experiments involve broad collaboration (particularly the large ones). I am happy that I was able to help facilitate this, both in my role at ORNL and on national committees. The Transport Task Force, which I helped set up, continues to encourage important cooperation in theory and computing, as well as in relating them to experiments.

Interestingly, the United States had better international than internal collaboration in previous years. Over the years, these activities have expanded through bilateral arrangements with other countries and multilaterally through ITER.

Joint work is also undertaken in the IFE area. I am fortunate to be chair of the Heavy Ion Beam Program Advisory Committee, undertaken as a virtual national laboratory by the Lawrence Berkeley, Lawrence Livermore, and Princeton laboratories. As we move into the period in which NIF will show energy gain in the laboratory, it will be important to strengthen cooperative activities within the IFE community, including benefiting from work in MFE.

One final set of comments: If people think that what we have accomplished to date was the most difficult part of the journey, they are in for a shock. It will be incredibly challenging to operate continuously (i.e., for weeks at a time) without failure. We have a long way to go in establishing components with the necessary reliability and maintainability. For that reason, I am not convinced that we have yet identified the optimum approach among our many options. Nevertheless, I remain convinced that fusion energy will be realized for the benefit of the world.

As to the attitude of fusion aficionados: Heck, any real one would have offered to teach the sheik's donkey two languages. *Hubieran ofrecido ensenarle español, tambien.*

Acronyms

APS	American Physical Society
ATF	advanced toroidal facility
BPX	burning plasma experiment
CEA	Commissariat à l'Énergie Atomique
CIT	compact ignition tokamak
COE	cost of electricity
CO_2	carbon dioxide
D	deuterium
DEMO	demonstration power plant (fusion)
DOE	Department of Energy (U.S.)
EBT	Elmo Bumpy Torus
EC	electron cyclotron
eV	electron volt
FESAC	Fusion Energy Sciences Advisory Committee (DOE)
GHz	gigahertz
H	hydrogen
He^3	helium-3
He^4	helium-4
HFIR	high flux isotope reactor
HV	high voltage
IAEA	International Atomic Energy Agency
ICSE	Intermediate Current Stability Experiment
IFE	inertial fusion energy
IFMIF	International Fusion Materials Irradiation Facility
INTOR	International Tokamak Reactor
ISX	Impurity studies experiment
ITER	international thermonuclear experimental reactor
JAERI	Japanese Atomic Energy Research Institute
JET	joint European torus
keV	kiloelectron volt
kV	kilovolt
LBNL	Lawrence Berkeley National Laboratory
LLNL	Lawrence Livermore National Laboratory
MA	megampere
MeV	megaelectron volt
MFE	magnetic fusion energy
MHD	magnetohydrodynamics
MIT	Massachusetts Institute of Technology
MW	Megawatt
n	Neutron

NACA	National Advisory Committee for Aeronautics
NET	next European torus
NIF	national ignition facility
NRL	Naval Research Laboratory
ORNL	Oak Ridge National Laboratory
OFE	Office of Fusion Energy
OFES	Office of Fusion Energy Sciences
ORO	Oak Ridge Operations (DOE)
p	proton
PPPL	Princeton Plasma Physics Laboratory
Q	fusion power produced/heating power to the plasma
RF	radio frequency
ST	spherical torus
SWAT	Shiva Winner Altruistic Trust
T	tritium
TAC	Technical Advisory Committee (ITER)
TEXT	Texas Tokamak
TFTR	Tokamak Fusion Test Reactor
TVA	Tennessee Valley Authority
UCLA	University of California—Los Angeles
UCSD	University of California San Diego
ZETA	Zero Energy Thermonuclear Assembly
UKAEA	United Kingdom Atomic Energy Authority

Glossary

Additional or auxiliary heating Heating auxiliary to ohmic heating, which decreases as the electron temperature increases. Examples include neutral beams, microwaves and radio frequency waves, and compression and shock waves.

Absolute zero The temperature zero on the Kelvin scale. Water freezes at 273 K and boils at 373 K. Hydrogen freezes at about 15 K and helium at 4 K.

Alcator C-Mod A high field tokamak at MIT, in Cambridge, Massachusetts.

Alpha particle The nucleus of a helium atom, consisting of two protons and two neutrons.

Anomalous transport The transport of heat and particles in excess of that due to collisional processes.

ASDEX A tokamak at the Max Planck Institute für Plasmaphysik, in Garching, Germany—stands for Axisymmetric Divertor Experiment.

Aspect ratio The ratio of the major radius to the minor radius of a toroidal plasma.

Beta The ratio of the plasma pressure to the pressure exerted by the magnetic field.

Blanket A component containing lithium placed outside a burning deuterium–tritium (D–T) plasma to capture fusion neutrons and generate more tritium.

Confinement time The time for energy or particles to leave the plasma.

Cryogenics The science that deals with the production of very low temperatures.

Current drive Inductive, using a transformer; or noninductive, using electromagnetic waves or particle beams.

Cyclotron frequency The frequency at which charged particles gyrate around a magnetic field.

DEMO A demonstration fusion power plant. The step before commercialization.

Density In plasmas, this term refers to the number density, i.e., the number of particles per unit volume.

Deuterium A stable isotope of hydrogen, whose nucleus contains one proton and one neutron.

Diagnostic An apparatus used to measure one or more plasma quantities, e.g., temperature and density.

Diffusion The random flow of heat (particles) down a thermal (density) gradient.

DIII-D A tokamak at General Atomics in San Diego, California.

Divertor A magnetic field configuration at the edge of the plasma that directs escaping plasma to plates that can handle the power. A pump is used to remove the gas that is produced.

Electron cyclotron heating Coupling electromagnetic waves into a plasma at the electron cyclotron frequency, usually using microwaves.

Energy breakeven A condition in which the fusion power produced equals the hearing power, denoted by $Q = 1$. Ignition corresponds to $Q = \infty$.

Electron volt (eV) The energy gained by an electron in passing through a potential of 1 V. When used as a measure of temperature, 1 eV = 10,600 K.

Fast ignition The use of a very-short-pulse laser beam (measured in picoseconds) to ignite a compressed fusion capsule in IFE.

Field lines and flux surfaces A field line is an imaginary line marking the direction of a force field. For magnetic fields, the field lines define flux surfaces, to which the charged particles are approximately constrained.

Fueling Supplying a plasma with more particles, e.g., by injecting gas or solid pellets of hydrogen.

Heliotron A stellarator in which one helical coil was originally wrapped around the torus. *See also* torsatron.

Heilotron-E An experiment at Kyoto University in Kyoto, Japan.

Hohlraum A cylindrical shell surrounding an IFE spherical capsule, containing the X-rays that compress and heat the capsule.

Ignition The condition under which the fusion power can support the plasma's heat losses.

Impurities Ions other than the basic plasma ion species, which are unwanted because they dilute the fuel and radiate heat out of the plasma, e.g., helium produced by fusion and materials blasted off the walls of the vacuum vessel.

Inertial confinement fusion (IFE) The use of powerful beams (lasers or particles) to implode a spherical capsule, containing deuterium−tritium (D−T) to such densities and temperatures that fusion occurs.

Instability of a plasma A situation in which control of the plasma or some element of it is lost. On a small scale, micro-instabilities lead to enhanced loss of particles and heat.

Ion cyclotron heating Coupling electromagnetic waves into a plasma at the ion cyclotron frequency, usually using radio frequency waves.

Isotope Each element is characterized by the number of protons in its nucleus. However, an element may have different numbers of neutrons in its nucleus, giving rise to isotopes. Hydrogen has one proton, while its isotopes, deuterium and tritium, have one and two neutrons, respectively. All have one electron.

Joint European Torus (JET) An undertaking of the European Union, this is the world's largest tokamak.

JT-60U This tokamak at JAERI's Naka site was the flagship of the Japanese tokamak program. It is being replaced by a superconducting version called the JT-60 Super Upgrade.

Large helical device (LHD) The world's largest stellarator. Of the heliotron-torsatron type, it is situated at the National Institute for Fusion Studies in Toki, Japan.

Lawson criterion This is the condition of *density × confinement time × temperature*, under which a plasma will generate more energy from fusion reactions than was required to create and sustain it. The level of the criterion depends upon the fusion reaction in question.

Magnetic confinement fusion (MFE) The use of magnetic field to contain a plasma while it is heated.

Major radius The distance from the center of a torus to the center of the plasma.

Minor radius The distance from the center of the plasma to the plasma edge.

Magnetohydrodynamics (MHD) A mathematical description of the plasma and magnetic field, which treats the plasma as an electrically conducting fluid.

National Ignition Facility (NIF) The world's largest laser system for inertial fusion energy (IFE) at the Lawrence Livermore National Laboratory in Livermore, California.

Neutron An elementary neutral particle.

Neutral beam A beam of high-energy neutral atoms that is injected into a plasma to heat it, e.g., 1 MW of deuterium atoms at 100 kV.

Noninductive current drive Driving currents in the plasma without the aid of a transformer, e.g., by injecting electromagnetic waves or particle beams.

NSTX A low-aspect-ratio tokamak at the Princeton Plasma Physics Laboratory, Princeton, New Jersey.

Ohmic heating The use of a current in the plasma to heat it.

Omega The second-largest operating laser for inertial fusion energy (IFE) at Rochester University, Rochester, New York.

Pellet injection Fueling a plasma by firing small pellets of frozen hydrogen at high velocity, e.g., 2-mm-diameter pellets at 1,000 m/s.

Plasma At a high enough temperature (generally greater than 10,000 K), electrons are stripped from atoms, producing a gas of free electrons and positive ions called a *plasma*.

Plasma equilibrium A condition in which the plasma is in force balance with a magnetic field.

Plasma parameters The physical quantities that characterize a plasma, such as density, temperature, pressure, and beta.

Poloidal direction The short way around a torus, usually denoted by the subscript p or θ.

Proton An elementary particle having a positive charge equivalent to the negative charge of the electron but possessing a mass approximately 1837 times as great.

Quark A fundamental matter particle that is a constituent of neutrons, protons, and other hadrons. There are six types of quarks.

Reverse field pinch (RFP) A toroidal magnetic confinement device in which the poloidal and toroidal magnetic field are of comparable value.

Radio frequency (RF) Frequencies generally in the range of 20 MHz to around 1 GHz. Electromagnetic waves at such frequencies are used to heat plasmas.

Shock wave A type of propagating disturbance characterized by an abrupt change in the properties of the medium in which it travels.

Stellarator A toroidal magnetic device in which the poloidal field is generated entirely by external coils, thereby eliminating the need for an induced plasma current.

Target In inertial fusion energy (IFE), the capsule containing a sphere of solid deuterium and tritium, either as a sphere on its own or as a sphere inside a holhraum.

Temperature A measure of thermal energy in units of degrees or electron volts.

Tokamak Fusion Test Reactor (TFTR) At the Princeton Plasma Physics Laboratory, this reactor was one of the only tokamaks to use deuterium and tritium plasmas.

Toroidal direction The long way around a torus, usually denoted by the subscript T or ϕ.

Toroidal field coils Coils that produce a magnetic field in the toroidal direction.

Torsatron A stellarator using two helical coils that wrap around the torus. *See also* heliotron.

Torus A geometrical shape that resembles a tire or a doughnut.

Transport The process by which particles and energy are lost to the plasma edge.

Tritium The heaviest isotope of hydrogen, with a nucleus consisting of one proton and two neutrons. It is radioactive and does not occur in abundance in nature.

Turbulence Randomly fluctuating, as opposed to coherent wave action, as seen when a wave breaks.

Wendelstein 7-X (W7-X) A large modular stellarator under construction in Greifswald, Germany.

CPSIA information can be obtained at www.ICGtesting.com
Printed in the USA
BVOW01*0615310713

327116BV00006B/83/P